LOGIC FOR COMPUTER SCIENCE

INTERNATIONAL COMPUTER SCIENCE SERIES

SELECTED TITLES IN THE SERIES

Theory of Computability: Programs, Machines, Effectiveness and Feasibility *R Sommerhalder and S C van Westrhenen*

Distributed Systems: Concepts and Design *G Coulouris and J Dollimore*

C Programming in a UNIX Environment *J Kay and R Kummerfeld*

An Introduction to Functional Programming through Lambda Calculus *G Michaelson*

Clausal Form Logic: An Introduction to the Logic of Computer Reasoning *T Richards*

Software Engineering (3rd Edn) *I Sommerville*

High-Level Languages and their Compilers *D Watson*

Programming in Ada (3rd Edn) *J G P Barnes*

Elements of Functional Programming *C Reade*

Interactive Computer Graphics: Functional, Procedural and Device-Level Methods *P Burger and D Gillies*

Common Lisp Programming for Artificial Intelligence *T Hasemer and J Domingue*

Program Derivation: The Development of Programs from Specifications *R G Dromey*

Object-Oriented Programming with Simula *B Kirkerud*

Parallel Processing: Principles and Practice *E V Krishnamurthy*

Real Time Systems and Their Programming Languages *A Burns and A Wellings*

Programming for Artificial Intelligence: Methods, Tools and Applications *W Kreutzer and B J McKenzie*

FORTRAN 77 Programming: With an Introduction to the Fortran 90 standard (2nd Edn) *T M R Ellis*

The Programming Process: An Introduction using VDM and Pascal *J T Latham, V J Bush and I D Cottam*

Prolog Programming for Artificial Intelligence (2nd Edn) *I Bratko*

Principles of Expert Systems *P Lucas and L van der Gaag*

Introduction to Expert Systems (2nd Edn) *P Jackson*

Computer Architecture *M De Blasi*

LOGIC FOR COMPUTER SCIENCE

Steve Reeves
Michael Clarke

QMW, University of London

ADDISON-WESLEY
PUBLISHING
COMPANY

Wokingham, England · Reading, Massachusetts · Menlo Park, California
New York · Don Mills, Ontario · Amsterdam · Bonn
Sydney · Singapore · Tokyo · Madrid · San Juan

Cover design by Crayon Design of Henley-on-Thames and printed by The Riverside Printing Co. (Reading) Ltd.
Typeset by Advanced Filmsetters (Glasgow) Ltd.
Printed and bound in Great Britain by The Bath Press, Avon.

First printed 1990.

British Library Cataloguing in Publication Data

Reeves, Steve
Logic for computer science.
1. Computer systems. Programming. Applications of mathematical logic
I. Title II. Clarke, Michael
005.131

ISBN 0–201–41643–3

Library of Congress Cataloging in Publication Data

Reeves, Steve
Logic for computer science/Steve Reeves and Michael Clarke.
p. cm.––(International computer science series)
Includes bibliographical references.
ISBN 0–201–41643–3
1. Logic, Symbolic and mathematical. I. Clarke, Michael
II. Title. III. Series.
QA9.R34 1990
004′.01′5113––dc20

90–32374
CIP

Preface

Aims

The aim of this book is to give students of computer science a working knowledge of the relevant parts of logic. It is not intended to be a review of applications of logic in computer science, neither is it primarily intended to be a first course in logic for students of mathematics or philosophy, although we believe that much of the material will be increasingly relevant to both of these groups as computational ideas pervade their syllabuses.

Most controversial perhaps will be our decision to include modal and intuitionistic logic in an introductory text, the inevitable cost being a rather more summary treatment of some aspects of classical predicate logic. We believe, however, that a glance at the wide variety of ways in which logic is used in computer science fully justifies this approach. Certainly classical predicate logic is the basic tool of sequential program verification, but modal and temporal logics are increasingly being used for distributed and concurrent systems and intuitionistic logic provides a basis for expressing specifications and deriving programs. Horn clause logic and resolution underlie the very widespread use of logic programming, while algorithms for automated theorem proving have long been of interest to computer scientists for both their intrinsic interest and the applications in artificial intelligence.

One major (and deliberate) omission is the standard development of the logical basis of set theory and arithmetic. These theories are so well covered in a number of excellent and widely available texts (many of which are referenced in the text or are among the sources we acknowledge at the end of this preface) that we preferred to use the space for less well-exposed topics. Of course, the need to formalize arithmetic and set theory has led to major developments in logic and computer science and we have tried to give the historical perspective, while referring readers elsewhere for the detail.

Different disciplines have different motivations for studying logic and correspondingly different conventions of notation and rigour. To keep the text within reasonable bounds we have decided to omit some of the lengthier explanations and proofs found in traditional logic texts in favour of introducing topics, hitherto considered more 'advanced', that are central to modern computer science. In many cases, where proof methods have previously been specified by non-deterministic sets of rules, we have been more precise than usual by giving algorithms and programs; in other cases we have relied on the background of our students to keep routine formal development to a minimum.

Another major departure is that we present many of the definitions and algorithms as computer programs in, not just one, but two programming languages. We have chosen Prolog and SML, partly because they are both highly succinct and suitable languages for the procedures we want to express, but also because they have their roots, respectively, in logic and the λ-calculus, two of the most important theoretical developments that underlie computer science and the theory of computability. Either of the languages is sufficient, but a student who carefully studies the programs in both languages will learn a lot about the theory and technique of declarative programming as well as about the logical definitions and algorithms that the programs express.

In Appendix A we give a brief introduction to the philosophy and facilities of both languages, but this is aimed to some extent at teachers and students with a considerable background in computer science. The less sophisticated will need access to one or other of the introductory texts that we recommend. That being said, the programs are designed to be readable and should relay some message even to non-programmers, perhaps helping them to realize that much of logic is inseparable from the notion of an effective algorithm, and even encouraging them to start programming.

Overall, our aim has been to show how computer science and logic are closely linked. We hope that students will see that what they might have considered a dry subject without obvious applications is being put to good use and vigorously developed by computer scientists.

Readership

Much of the material has been tested in a course given to first-year undergraduate students in computer science who, at that stage, have had an introductory course in discrete mathematics and a first programming course that emphasizes recursion, inductive proof and scope of definition. So they already have a fair grasp at a semiformal level of notions such as set, function, relation, formal language, free and bound variables and mathematical induction, and we believe that such a background will, if not already standard, soon become so for first-year students of computer science. The students, we are glad to say, bear out our conviction that an introductory logic course can successfully go beyond what is usually considered to be the appropriate level. They are able actually to do proofs using the methods we teach and are surprised and challenged by the idea of several logics. We feel that this is because computer science, properly taught, makes the study of logic easier, and vice versa. The activity of constructing and reasoning about programs is not all that different from the activity of constructing and reasoning about proofs.

Acknowledgements

Our colleagues, also, have made a big contribution to the development of the course, and subsequently the book. We would single out for special mention, in no particular order, Peter Burton, Wilfrid Hodges, Doug Goldson, Peter Landin, Sarah Lloyd-Jones, Mike Hopkins, Keith Clarke, Richard Bornat, Steve Sommerville, Dave Saunders, Mel Slater, John Bell and Mark Christian, as well as those further away – Alan Bundy, Dov Gabbay – who have influenced our views on the more advanced topics.

Finally, it will be obvious that we have been strongly influenced, and greatly helped, by many other texts whose development of the subject we have studied, and in many instances borrowed. These have included Hodges (1977), *Logic*, Hamilton (1978), *Logic for Mathematicians*, Boolos and Jeffrey (1980), *Computability and Logic*, Scott *et al.* (1981), *Notes on the Formalization of Logic*, Lloyd (1984), *Foundations of Logic Programming*, and Martin-Löf (1985), *Constructive Mathematics and Computer Programming*.

Steve Reeves
Mike Clarke

QMW, University of London
November, 1989

Contents

1 Introduction

1.1 Aims and objectives

This book will differ from most others with similar titles because we aim to give you not one or two ways of looking at logic, but many. The forms of reasoning that are fundamental to computer science are not necessarily those most familiar from a study of mathematics and this gives us the opportunity to develop the subject along two dimensions, looking not only at different methods for implementing one particular mode of reasoning, but also at different ways of formalizing the process of reasoning itself.

There are many reasons why a computer scientist should need to study logic. Not only has it historically formed the roots of computer science, both Church's and Turing's work being motivated by the decision problem for first-order logic, but nowadays we are finding conversely that computer science is generating an explosion of interest in logic, with the desire to automate reasoning and the necessity to prove programs correct.

Basically, logic is about formalizing language and reasoning, and computer science addresses similar problems with the extra task, having formalized them, of *expressing* those formalizations, in the technical sense of producing mechanisms which follow the rules that they lay down. This, indeed, has led to the recent use of computer science for investigating logics in an experimental way, exploring some of them much more thoroughly than was possible when the 'computer' was a person rather than a machine.

What we hope then to show is that computer science has grown out of logic. It is helping to suggest new ideas for logical analysis and these logical ideas are, in turn, allowing computer science to develop further. The two subjects have each contributed to the growth of the other and still are contributing, and in combination they form an exciting and rapidly growing field of study.

1.2 Background history

In the middle of the last century Boole laid down what we now see as the mathematical basis for computer hardware and propositional logic, but the logics that we are going to look at really started towards the end of the century with the work of Gottlob Frege, a German mathematician working in relative obscurity. Frege aimed to derive all of mathematics from logical principles, in other words pure reason, together with some self-evident truths about sets (such as 'sets are identical if they have the same members' or 'every property determines a set'). In doing this he introduced new notation and language which forms the basis of the work that we shall be covering. Until Boole and Frege, logic had not fundamentally changed since Aristotle.

Frege's huge work was (terminally) criticized at its foundations by Bertrand Russell who found a basic flaw in it stemming from one of the 'self-evident' truths upon which the whole enterprise was based. However, Russell developed the work further by suggesting ways of repairing the damage. He also introduced Frege's work to the English-speaking mathematicians since not many of them, at that time, read German. Russell, who did read German, saw that the work was important and so publicized it.

1.3 Background terminology

We are going to be doing what is usually known as 'mathematical logic' or 'symbolic logic' or 'formal logic'. That is, we are going to use ordinary, but careful, mathematical methods to study a branch of mathematics called logic. Before we start to look at what logic actually is we shall try to make the context in which we are working a bit clearer. To make the discussion concrete we can think in terms of the typical introductory programming course that you may have followed.

Such a programming course not only teaches you how to use the constructs of the language to produce the effects that you want when the program is executed, but it also teaches you the distinction between the language that you write programs in and the meaning of the statements of that language in terms of the effect that they have when executed by a computer. If the course was a good one, it will also have taught you how to reason about programs – perhaps to show that two apparently different programs are equivalent. Logic is the study of formal (that is symbolic) systems of reasoning and of methods of attaching meaning to them. So there are strong parallels between formal computer science and logic. Both involve the study of formal systems and ways of giving them meaning (semantics). However, in logic you study a wider variety of formal systems than you do in computer science, so wide and so fundamental that logic is used not only as one of the mathematical tools for studying programming, but also as a foundation for mathematics itself. This ought to set the alarm bells ringing, because we have already said that we were going to use mathematics to study logic, so there is an apparent circularity here. It is certainly the case that circular or 'self-referential' discussion such as this is very easy to get wrong but the notion of self-reference is a central one in computer science and, in fact, is exploited rather than avoided.

In logic we deal with the issue by putting the logic we are going to **study** in one compartment and the logic we are going to do the studying **with** in another. These compartments are realized by using different **languages**. The logic that is the object of our study will be expressed in one particular language that we call the **object language**. Our study of this logic and language is carried out in another language which we call the **observer's language** (you might also see the word **meta-language** for this).

The idea should already be familiar to you from studying foreign or ancient languages. In this case Latin, for example, might be the object language and your native language, in which you might have discussed the details of Latin syntax or the meaning of particular Latin sentences, is the observer's language. In mathematics, the symbolisms of calculus, set theory, graph theory and so on, provide the object language and again your native language, augmented perhaps with some specialized mathematical vocabulary, is used as the observer's language. In programming, the object language is a 'programming' language such as Pascal, Lisp or Miranda and the observer's language is again your native language augmented with the appropriate mathematical and operational notions.

EXAMPLE 1.1

Consider the statement

```
times 0 do print * od = donothing
```

This, in fact, is a statement in the observer's language about the equivalence of two statements in one of our local programming

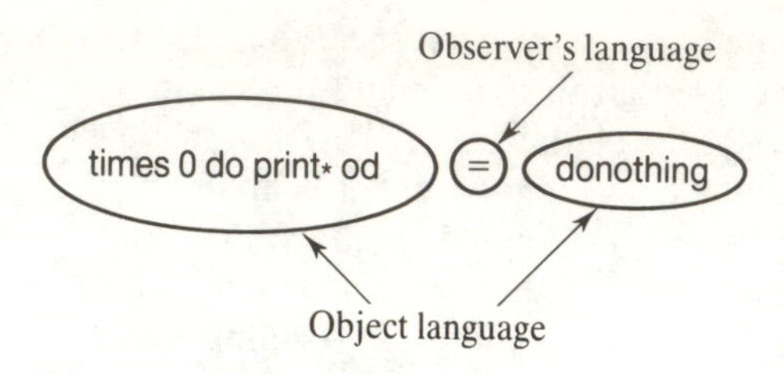

Figure 1.1.

languages. Although you may have guessed correctly, you have no means of saying with certainty which are the symbols of the object language and which are symbols of the observer's language until the object language has been defined for you. In fact, the distinction is as shown in Figure 1.1.

Exercise 1.1 Now you are invited to use your linguistic, mathematical and programming experience to do a similar analysis of the following statements into observer and object languages.

(a) The sentence 'They feeds the cat' is ungrammatical in English.

(b) The French translation of the English phrase 'Thank you very much' is 'Merci beaucoup'.

(c) The equation $E = mc^2$ holds in special relativity.

(d) There is no real value of x that satisfies $x^2 - 2x + 2 = 0$.

(e) There is no real value of foo that satisfies $x^2 - 2x + 2 = 0$.

(f) If $x^2 - 2x + 1 = 0$ then $x = 1$.

(g) If $x^2 - 2x + 1 = 0$ then x must be 1.

(h) If $x^2 - 2x + 1 = 0$ then x must be unity.

(i) '$E = mc^2$ holds in special relativity' cannot be proved.

(j) The statements $x := x + 1$; $x := x + 1$; are equivalent to $x := x + 2$; in Pascal.

(k) '**if** ... **then** ... **else**' is a statement form in Pascal.

You probably found that one or two of these exercises were borderline cases and caused you to point out, more or less forcibly, that it would be a lot easier if you had an actual definition of the object language in front of you. This is the first thing we do when we embark on the development of propositional logic in Chapter 2.

1.4 Propositions, beliefs and declarative sentences

The basic items that logic deals with are **propositions**. Philosophers have given a variety of answers to the question 'What is a proposition?' but since we are dealing with the mathematics rather than the philosophy of logic it does not really matter for our purposes. One answer, however, is that a proposition is what is common to a set of declarative sentences in your native language that all say the same thing. Philosophers then have to argue about what 'all say the same thing' means, but fortunately we do not.

Propositions communicate judgements or beliefs and since beliefs are themselves manifested as states of mind (it is hard to see what else they could be) the act of believing or the stating of propositions allows, with practice, the representation in your mind of complex objects both physical and abstract. We seem to be a long way from the limits of the human race, as a whole, in representing things mentally, and reasoning with them, and we are in the Stone Age when it comes to building ourselves tools for doing so. This is why the study of formal methods of manipulating propositions, logic in other words, is so important.

Since computer science is one discipline in which the objects that we want to reason about are extraordinarily complex, and often abstract and purely formal, the need for logic here is especially clear.

Of course, the fact that beliefs are states of mind means that we cannot directly manipulate them; neither can we manipulate propositions, since they are expressions of those states of mind. What we do is to manipulate sentences in some language which map on to propositions and beliefs.

The language with which we, the authors, are most familiar for this task is the natural language called 'English'. We use it to express our beliefs as propositions, for the purpose of transferring them to each other, testing them and so on. When one of us says to the other 'I believe that Moto Guzzi manufacture the best motorcycles in the world' he conveys part of his current state of mind, in particular a part that expresses a certain relation between himself and motorcycles.

In general, then, we use English to express our beliefs. However, we need to refine this statement since English is rather complicated and not all of English is used for this purpose. There are only certain sentences in English that convey beliefs, that is, express propositions, and these are the **declarative** sentences.

> **Definition 1.1:** A **declarative sentence** is a grammatically correct English sentence that can be put in place of '...' in the sentence 'Is it true that ...?' with the effect that the resulting sentence is a grammatically correct English question.

One might expect further restrictions here, though. The definition has a rather syntactic bias to it, and English is notoriously expressive. We cannot go into it fully here, but a good introductory discussion can be found in Hodges (1977).

Exercise 1.2 Decide whether the following are declarative sentences or not:

(a) What is your name?

(b) Close the door!

(c) Grass is green.

(d) Grass is red.

(e) It is wrong.

(f) I am honest.

(g) You must not cheat.

(h) It is false that grass is red.

1.5 Contradictions

By this stage you should have some feel for how beliefs are manipulated in natural language. But how are beliefs actually useful? What is their reason for existing? Basically beliefs give a description of the world as it is or might be. For example, if I have a system of beliefs about the laws of mechanics (a description of part of the world) I can generate beliefs about the solar system without actually having to go out there and make measurements. If I have a system of beliefs about my friends, I can predict their behaviour in certain situations without the possible embarrassment of engineering and being in those situations. Again, I can have a set of beliefs about numbers and reason about the result of the sum $2 + 2$, without actually creating two different sets of cardinality 2, amalgamating and counting them.

So systems of belief allow decisions to be made, facts to be conjectured; it seems that they can do anything. However, there is one limitation. You cannot simultaneously hold two different beliefs which you know contradict one another. Since we have said that logic is important because it allows us to manipulate beliefs, it follows that a fundamental task of logic is to be able to decide whether or not a set of beliefs is contradictory. In simple cases, as we shall see, logic can do this in a mechanical way. But there are inherent limitations and it may be that, ultimately, even the most ingeniously programmed machine cannot do as well at manipulating propositions as the most careful person. This belief has not yet been contradicted.

1.6 Formalization

Formalization is the process of constructing an object language together with rules for manipulating sentences in the language. One aim in doing this is to promote clarity of thought and to eliminate mistakes.

Another equally important issue, one that gives rise to the term 'formalization' itself, is that we provide a means of manipulating objects of interest without having to understand what we are doing. This sounds at first like a retrograde step, but let us give an example: arithmetic arose out of the formalization of counting. So we now have a set of rules which we can follow to add together numbers correctly without needing to understand what numbers are or what adding up is. Of course, we can always go back to the original physical act and see that adding up comes from the process of counting up to a particular number n with reference to one group of objects and then starting from $n + 1$ in counting a second group. The answer in this case is what we would formally call the sum of the numbers concretely represented by each group.

So the power of formalization is that, once formalized, an area of interest can be worked in without understanding. If the agent following the rules is a human being this might be a mixed blessing, since understanding at the intellectual level is a strong motivation for getting things done. But, if you want to write a computer program to reason, then formalization is essential. Equally essential, if the results are to be useful, is to be able to prove that, as long as the rules are correctly applied, the results will be correct.

For instance, a programmer employed in the financial sector may have, in the form of a set of beliefs that are related in complicated ways, an idea of how the stock exchange works. It is the abstract structure of these relationships which models the concrete structure of the stock exchange and forms a model of how the stock exchange works. The programmer will then formalize this model when writing a computer system to automatically deal in the stock exchange, say. Now, if you look at the program, it is clear that the names of the objects in the program do not matter; nor does the language in which they are written. What matters is that the relationships in the real thing are faithfully and fully represented in the program. This is the sense of formalization that we are concerned with: the program should model the form of the real thing in interaction between its parts.

In the next chapter we make a start by looking at a simple form of declarative sentence and we show how it can be used to formalize some basic instances of reasoning.

SUMMARY

- Mathematical logic began towards the end of the last century when Frege developed what is now the predicate calculus.

- Mathematical logic involves applying standard mathematical methods to the study of systems that themselves can be used to formalize mathematics. The apparent circularity is overcome by distinguishing between the *object* language, in which the formal system is expressed, and the *observer's* language in which properties of the formal system are expressed and reasoned about.
- The basic items that logic deals with are *propositions*. Propositions are used to express beliefs. In natural language they are represented by *declarative sentences*.
- The notion of belief is a very general one; nevertheless there are some restrictions on the way beliefs can be manipulated in mental reasoning. For example, you *cannot simultaneously hold contradictory beliefs* (without at least being aware that something is wrong).
- The importance of formalization is that once a particular area of mathematics or computer science has been formalized, reasoning in it can be carried out purely by symbol manipulation, without reference to meaning or understanding, and mathematical properties of the reasoning process can be clearly stated and proved.

2 Formalizing the Language

2.1 Informal propositional calculus

This chapter introduces a language which is very widely used in mathematics and computer science for expressing sets of propositions and calculating whether they are consistent, in other words contain no contradictions.

2.1.1 The language

We shall use $\rightarrow$, $\wedge$, $\vee$, $\neg$ and $\leftrightarrow$ as our standard symbols for the **connectives**. These symbols form part of the **alphabet** of the language of propositional logic. Other elements of the alphabet are the parentheses,) and (, and a set of propositional variables, for instance $\{p, q, r, s\}$. We can now give a proper definition of the conditions that a sequence (string) of symbols must satisfy to be a sentence of propositional logic.

Definition 2.1: If P is a set of propositional variables then:

(1) a propositional variable from the set P is a sentence;

(2) If $\mathscr{S}$ and $\mathscr{T}$ are sentences then so are $(\neg\mathscr{S})$, $(\mathscr{S} \wedge \mathscr{T})$, $(\mathscr{S} \vee \mathscr{T})$, $(\mathscr{S} \rightarrow \mathscr{T})$ and $(\mathscr{S} \leftrightarrow \mathscr{T})$;

(3) no other sequences are sentences.

We can, for example, use the definition to show that p, $(p \rightarrow q)$, $(p \wedge (\neg q))$ and $((p \wedge q) \rightarrow r)$ are sentences, because p is a sentence by clause (1), since p is a propositional variable, $(p \rightarrow q)$ is a sentence because p and q are sentences by clause (1) and hence, by the fourth condition in clause (2), the whole string of symbols is a sentence. The other cases follow similarly.

In practice, to keep the number of parentheses to a minimum, there is a convention that $\neg$ takes precedence over or, as it is sometimes put, 'binds more tightly' than $\wedge$ and $\vee$, which in turn bind more tightly than $\rightarrow$ and $\leftrightarrow$. Also, outside parentheses can often be omitted without ambiguity. So $(p \vee (\neg q))$ would usually be written as $p \wedge \neg q$ and $((p \wedge q) \rightarrow r)$ as $p \wedge q \rightarrow r$.

Furthermore, we can see that $\neg p)$, for instance, is not a sentence since, although we have that p is a sentence (clause (1)), none of the other clauses makes $\neg p)$ a sentence so by clause (3) it is not a sentence.

The set of symbol sequences (strings) defined in this way is called the **set of (well-formed) sentences** or **language** of propositional logic. The form of the definition is important not only because it is the first of many that we shall be seeing, but also because it determines the property of 'being a sentence of propositional logic' as being **decidable**. That is, the question 'is this sequence of symbols, formed from the alphabet of propositional logic, a sentence of propositional logic?' can always be answered correctly either 'yes' or 'no'. Later on we shall see some similarly structured questions which cannot be answered in all cases. These will be called **undecidable** questions.

There are two ways to be sure that this (or any) definition gives rise to a decidable property; either you can construct a proof that it is so or you can construct a program which always gives the correct answer to the question 'is this string a sentence?'. By 'always gives the correct answer' here we mean that, whenever the question is asked, the program answers it correctly before it terminates – and it always terminates. Clearly, to implement the definition as a program involves much more work than only proving that the definition gives a decidable property, but for the extra work we gain a program that can always answer the question correctly without further work on our part.

With a suitably expressive programming language we can use the above definition, of sentences based on the set P of propositional variables, to give us almost directly a programmed decision procedure. First, though, we have to represent the forms of sentence defined by the grammar in the language that we will use.

In this book we use two programming languages, SML and Prolog, that are becoming widely used in computer science for implementing algebraic and logically based calculations. These languages are also theoretically interesting in their own right. SML is based on the λ-calculus and type inference, while Prolog is based on the notion of logic programming (see Chapter 7). An introduction to each of the languages is given in Appendix A. Using the datatype feature of SML makes it easy to define all the possible forms of sentence. For instance, just as the definition above uses the phrase '... if $\mathcal{S}$ is a sentence then so are $(\neg\mathcal{S})$ and ...' then we can use the phrase '... S = Not of S | ...' in SML to represent the same idea. Notice that instead of using the symbol '$\neg$' we have used the symbol 'Not'; this is done so that later on when we write programs to manipulate these sentences we do not get too many, possibly confusing, sequences of symbols, that is we prefer the words.

So, using datatypes the type SENT, which represents sentences of the language, can be defined in SML as:

```
datatype   SENT = Prop of string
                | Not of SENT
                | And of (SENT * SENT)
                | Or of (SENT * SENT)
                | Imp of (SENT * SENT)
                | Eq of (SENT * SENT);
```

This means that a sentence, that is a value of type SENT, is either of the form Prop ("p") or Not (S) where S is a value of type SENT, or And (S,T), where S and T are both values of type SENT, and so on. The way in which this declaration represents sentences as defined above should now be clear.

Of course, writing down sentences using the form described in this declaration is not as convenient for us as the form in the original definition since there is more to write. But we must remember that the language SML is fixed (and completely formal, of course) whereas we, as humans, have a certain amount of flexibility when it comes to adapting ourselves to other notations. So, since the programming language will not adapt, we have to.

However, it is usual to mitigate the somewhat verbose programming representation of the components of a language when, say, non-specialists have to use it by putting a piece of software called a parser between the user and the program being used. The task of the parser is to take a string of symbols from the user and, for those that are suitable, to transform them into the required form from the programmer's point of view. Thus, we would, in any suite of programs to do things as yet unspecified with logical statements, expect to have to write a program which carried out this task. It turns out that to do this task properly is not a trivial thing. However, parsers are such important and ubiquitous pieces of software that a lot of effort has gone into making their design fairly routine. In the SML programs that we write in the rest of the text we will be using just the internal form of the language, but in Appendix B we have given the text of a parser so that the interested reader may see what is required to implement such a program. It is not, however, necessary to understand how the parser works, or even to acknowledge its existence, in order to understand what follows, so the uninterested reader need not worry. (In fact, by defining the constructor symbols to be infixed, we can circumvent the need for a parser somewhat, which is just what we do below in Prolog, but since this was a good place to bring to the reader's notice the idea of parsing we decided on the current presentation.)

Having represented the sentences we can now go on to describe how to do computations with them. First, consider the problem of deciding whether a string is a sentence based on a set of propositional variables P, as above. The following is an expression of the definition above but given in SML:

```
fun decide P (Prop v) = v memberof P
  | decide P (Not(l)) = decide P l
  | decide P (And(l,r)) = (decide P l) andalso (decide P r)
  | decide P (Or(l,r)) = (decide P l) andalso (decide P r)
  | decide P (Imp(l,r)) = (decide P l) andalso (decide P r)
  | decide P (Eq(l,r)) = (decide P l) andalso (decide P r);
```

We can see that the program always terminates since it either returns a value (true or false) immediately or, if a recursive call is made, the arguments of the recursive call are strictly smaller, as pieces of text, than the preceding case. Therefore, given any finitely long argument, there can only be a finite number of recursions, so the program will always terminate. The correctness of its answer follows from the closeness of the program to the original definition. We conclude, therefore, that the property of being a sentence of propositional logic is decidable.

2.1.2 The Prolog version of 'decide' and the enumeration of sentences

In Prolog (and again see Appendix A for details of the language) the decision procedure can be expressed as

```
:- op(510, fx,    [~]).
:- op(520,xfy,    [/\]).
:- op(530,xfy,    [\/]).
:- op(540,xfx,  [->]).
:- op(550,xfx,[<->]).
member(X,[Y|Z]):- X=Y; member(X,Z).
decide(  S):- member(S,[p,q,r,s]).
decide(~S):- decide(S).
decide(  S):- (S=Q/\R;S=Q\/R;S=Q->R;S=Q<->R),
              decide(Q),decide(R).
```

The major difference between the SML and Prolog versions is that with Prolog we define a relation rather than a function, but it should nevertheless be apparent that the programs are essentially the same operationally and what was said above about the recursive calls, termination and decidability applies equally here.

Prolog allows you to define your own infixed operators and this program is complete and can be used as it stands, without the need for an additional parser. We have taken the opportunity to invent some symbols for the logical connectives that are as close as the normal keyboard will allow to the symbols for the logical connectives that we use in this book. The first five lines of the program are system-defined predicates for specifying the precedence and associativity of the operators and you will find an explanation of this in Appendix A or any good book on Prolog.

Prolog programs are used by expressing a request for a computation as a **goal** (again, see Appendix A for more practical, and Chapter 7 for more theoretical, detail). For example, to use the program above to check whether or not the string of symbols **p/\q** -> **r** is a sentence of the language, you type

```
?- decide(p/\q -> r).
```

to which the Prolog interpreter would output '**yes**', whereas for

```
?- decide(p + q).
```

you would get the output '**no**', and for

```
?- decide (-> p).
```

you would be told that there is a syntax error because the connective -> is defined to take two arguments.

As well as the question of decidability another important property is that the sentences of propositional logic are **enumerable**, that is all the sentences that get the answer 'yes' can be arranged in a list with a first member, a second member, and so on. With a little experience of Prolog one might think that the backtracking mechanism could be used to generate this list by repeatedly satisfying the goal

```
?- decide(S).
```

However, this does not work because of the fixed-order, depth-first nature of Prolog's execution model. With two propositional variables **p** and **q** you would, in fact, get the output

```
p;
q;
~p;
~q;
~~p;
~~q;
~~~p;
```

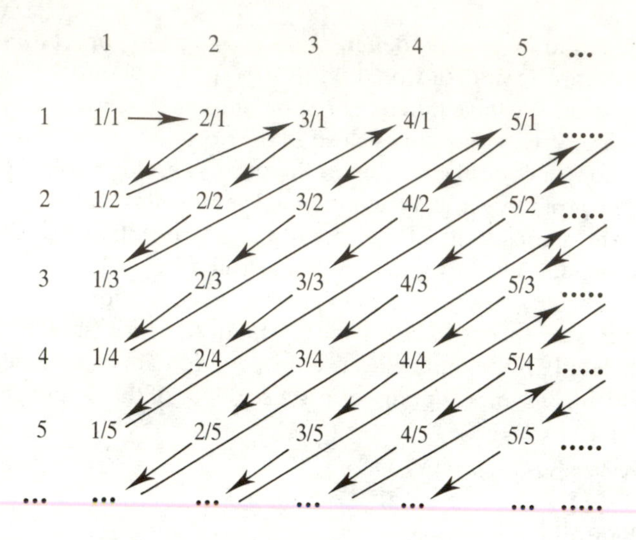

Figure 2.1.

and so on *ad infinitum*. Because the program clauses dealing with ~ come before those dealing with the other connectives, a formula such as **p/\q** can never be generated. It is clear that simply reordering the clauses will not help either.

One way to get round this is to use an idea invented around 1880 by Georg Cantor. Cantor wanted to find out whether all infinite sets are the same size (they are not in fact), and he defined a set as **enumerably infinite** if it could be put into one-to-one correspondence with the positive integers. He noticed that a similar problem to the one we have encountered here arises with the rationals, that is numbers of the form i/j where i and j are positive integers. If you start listing them in order 1/1, 2/1, 3/1, ... you will never, for example, get to 1/2. Cantor's way round this was to imagine the rationals arranged in a two-dimensional array in which i/j is in the ith column and jth row (see Figure 2.1).

The idea is to enumerate the elements of the array by starting in the top left-hand corner with 1/1 and following the path shown to give the list 1/1, 2/1, 1/2, 3/1, 2/2, 1/3, 4/1, 3/2, ... It is intuitively clear that by doing this you eventually get to any point i/j. If intuition does not satisfy you, as indeed it should not if you are going to study logic, then it is straightforward by elementary algebra to show that i/j is in position $(i+j-2)(i+j-1)/2+j$ in the list. Conversely, we can invert the mapping to show that the nth rational in the list is i/j where $j = n-(k-2)(k-1)/2$, $i = k-j$ and $k = \left\lfloor (3+\sqrt{8n-7})/2 \right\rfloor$.

We can use the same idea to enumerate logical formulas. To keep things simple for the moment suppose we have only one connective, → say.

	p	$p \to p$	q	$(p \to p) \to p$	r
p	$p \to p$	$(p \to p) \to p$	$q \to p$	$((p \to p) \to p) \to p$	...
$p \to p$	$p \to (p \to p)$	$(p \to p) \to (p \to p)$	$q \to (p \to p)$	...	...
q	$p \to q$	$(p \to p) \to q$	$q \to q$	...	...
$(p \to p) \to p$	$p \to ((p \to p) \to p)$	...	...	...	...

Figure 2.2.

Then, following the recursive scheme given above, we know that any formula is either a sentence letter or of the form $(\mathscr{S} \to \mathscr{T})$ where $\mathscr{S}$ and $\mathscr{T}$ are both formulas. Now we do exactly the same as for the rationals. $(\mathscr{S} \to \mathscr{T})$ is in the column corresponding to $\mathscr{S}$ and the row corresponding to $\mathscr{T}$.

The propositional variables count as formulas, of course. If they are finite in number then they can be listed before all the compound formulas. If we have an infinite supply of propositional variables then there is a slight complication because we must ensure that any given one is eventually brought into use. One way round this is to place the propositional variables at odd numbers in the list, with compound formulas at the even-numbered positions. The top left-hand corner of the array of compound formulas would then, for propositional variables $p, q, r, \ldots$, be as in Figure 2.2 (we have as usual omitted the outermost pair of parentheses).

We can now use exactly the same scheme of enumeration as we did for the rationals to generate the list of formulas $p \to p, (p \to p) \to p, p \to (p \to p), \ldots$ which, when merged with the sentence letters, will be at positions $2, 4, 6, \ldots$. We can imagine this mapping to be a code, a function c say, that takes formulas to integers. Again there is an inverse decoding function, d say, that takes integers to formulas. As above it is straightforward to show that

$$c(\mathscr{S} \to \mathscr{T}) = (c(\mathscr{S}) + c(\mathscr{T}) - 2)(c(\mathscr{S}) + c(\mathscr{T}) - 1) + 2c(\mathscr{T})$$

and, for n even, $d(n)$ is $d(i) \to d(j)$ where $j = [n - (k-2)(k-1)]/2$, $i = k - j$ and $k = \left\lfloor (3 + \sqrt{4n - 7})/2 \right\rfloor$.

We can then write these functions as Prolog predicates and, by enumerating the integers in increasing order and decoding them into formulas, we can realize our original goal of demonstrating an effective procedure for enumerating all the sentences of the language – or in practical terms, as many as we want.

```
:- op(540,xfx,[->]).
int(1).
int(K) :- int(J), K is J + 1.
split(N,I,J) :- int(K),L is K * (K - 1),L >= N,!,J is (N - (K - 2) * (K - 1))//2,
                I is K - J.
```

```
code(S,N) :- atom(S),name(S,[I]),N is 2 * I - 193.
code(S -> T,N) :- code(S,I),code(T,J),N is (I+J-2) * (I+J-1)+2 * J.

decode(N,S) :- 1 is N mod 2,!,I is (N+193)//2,name(S,[I]).
decode(N,S -> T) :- split(N,I,J), decode(I,S), decode(J,T).

enumerate(N) :- int(K),L is 2 * K,decode(L,X),write(X),nl,K >= N.
```

This is a complete program for coding, decoding and enumerating sentences that will work as it stands. It will be seen that it includes some features of Prolog that are of no logical interest, but are necessary to manipulate numbers and symbols at a basic level. The **name** predicate for example is a system utility for splitting an atom into a list of characters. It is used here to generate propositional variables starting at the letter 'a' (193 being twice the internal numerical code for 'a' minus one). The **split** predicate does the square root computation by a method of testing that uses only integer arithmetic. It could be replaced in most Prolog systems by the system's square root operator.

Of course enumerating all possible sentences is not much practical use but, in the theoretical study of what the limits to computation are, it is important to show that you can in principle carry out such an enumeration. Furthermore, the idea that you can code formulas using arithmetic is an important one in its own right, quite apart from its use here for enumeration. In one of the most significant results in the history of mathematics Gödel used the idea to show, in 1931, that any logical basis for arithmetic is bound to be inadequate in a certain technical sense. We say a little more about this later.

Exercise 2.1

(a) Work out how you might modify the Prolog program for enumerating sentences with -> so that it could handle the complete set of connectives ~, /\, \/, ->, <->.

(b) Calculate by hand what the 1000th compound sentence is, that is what $d(2000)$ is.

(c) Write a similar program for enumerating sentences in SML.

2.1.3 Giving meaning to the language

So far we have defined the **form** of the language that we are studying. We can now decide whether or not a string of symbols from the alphabet is a well-formed sentence of this language. However, we do not yet have a way of giving any **meaning** or **semantics** to these sentences. It is as if you had been given a manual for Pascal with the part that gives the meaning, usually a set of rules that tell you how a given statement is executed (known as **operational semantics**), omitted from the manual. You would be in a position to write programs that were correctly formed, as far as the Pascal syntax is concerned, but you would not know what the program denoted or what its effect would be. You would not know its meaning.

At this point we are in a similar position with propositional logic; we can write well-formed sentences but we cannot say what they mean or relate them to any other language.

To give a meaning to a language we first have to associate the sentences of the language with some class of objects. For instance, in English, we begin our acquisition of language with simple utterances that consist of single words: 'Dog!', 'Cat!', 'Car!' and so on. Our understanding of the meaning of these words is judged by our use of them. That is, we say that someone understands the word 'cat' if they point to a cat as they say it. If someone points to a cat and says 'Dog!' then we conclude that they do not understand the meaning of the word.

At this simple level, at least, we can imagine that there is a mapping between the words and the objects that they denote. Then, as our natural language develops, we understand words for verbs, adjectives, relations and so on. Again, our understanding is demonstrated by our use of the words in certain situations. Of course, a natural language like English is far more complicated than any artificial language (such as a programming language or a logical language) and, although our methods will allow us to give a semantics to the languages that we shall study in this book, it is not yet, and may never be, possible to give a semantics to English in the same manner (if only because the language is not static).

So we have first to decide what objects, perhaps in an abstract sense, we are to have our sentences denoting. For the moment we are going to make an assumption that will mean our semantics is **classical**. The assumption is that any sentence is – has for its meaning something that is – either true or false. To make the meanings of 'true' and 'false' absolutely clear we can do what we do when telling someone the meaning of any other word – we can point to an instance. In this case, to explain our usage of 'true' and 'false', we refer you to statements that we can confidently assume you will recognize as true and false. Every reader can be assumed to know that the numbers one and zero are not equal. So we can say that what a sentence refers to will either be the same thing that the arithmetic statement '$0 = 0$' refers to or it will be the same thing that the arithmetic statement '$0 = 1$' refers to. Here we are taking a view of semantics that is called **denotational** – we say what each sentence in the language denotes. Contrast this with what was said earlier about Pascal, where we spoke of the semantics of the language being specified by the actions that can be attributed to a particular sentence of the language, known as **operational** semantics. As it happens, Pascal is a simple enough language for it to be given a denotational semantics too, although quite a lot of work had to be done before the underlying mathematics was properly worked out.

We could contemplate giving an operational semantics to our propositional logic in which, perhaps, one action was performed and called 'true' and another was performed and called 'false' depending on whether the result of testing a value in a computer register is what we conventionally call true or false. It turns out, however, that such an operational approach is too cumbersome when all we really want to work with are abstract notions such as truth and falsity. The idea of letting truth reside in a particular action involves all the detail of how the action is carried out, how long it takes and what happens if it goes wrong. For other purposes, particularly those involving getting an effect when you execute a program, such as an invoice being printed or a line drawn on a screen, the operational detail is vital; but here it just complicates matters, so we take the denotational approach.

Every sentence of the language is either true or false, so we need rules for calculating which of these two values a given sentence has. The situation here is similar to the usual algebra of arithmetic where we have operations such as multiplication and addition, symbolized by $\times$ and $+$, and letters such as x and y which are understood to stand in place of numbers. In propositional logic we have five operation symbols and a quantity of sentence variables that can stand in place of sentences. Just as we can give a meaning to an algebraic statement in terms of the meanings, or denotations, of the operation symbols, which operate on numbers, so we can give a meaning to the logical operators, which here we call connectives, by saying how they act on the denotations of sentences, that is on truth and falsity.

2.2 Arguments

2.2.1 Informal arguments

A notion that is central to our work is that of **argument**. The language that we introduced above will allow us to express, and show valid, arguments such as

If Socrates is a man then Socrates is mortal.

Socrates is a man.

Therefore, Socrates is mortal.

It should not allow us to show valid the argument

Socrates is a man.

Therefore, Socrates is mortal.

even though, by knowing the meaning of the words, we would say that the conclusion holds. If the textual form of the argument is changed to

If Ssss is an xxxx then Ssss is a yyyy.

Ssss is an xxxx.

Therefore Ssss is a yyyy.

we can guarantee that you cannot make an assessment of the argument from its meaning, yet we anticipate that you will accept it as valid by virtue of its structure.

So we are going to judge the validity of arguments by their **form**, not their meaning. This means that even the following argument is valid:

Paris is in Australia and Australia is below the equator.

Therefore, Paris is in Australia.

because its form is that of a valid argument, even though your geographical knowledge tells you the conclusion is false. Also, the following argument is not valid despite all the statements, including the conclusion, being true:

The Eiffel Tower is in Paris or Paris is in France.

Therefore, the Eiffel Tower is in Paris.

So, the correctness of an argument is not governed by its content, or its meaning, but by its logical form. We are going to make this idea much more precise and at the same time abstract away from the particular statements involved in the arguments, which makes the form much clearer.

Exercise 2.2

(a) Say whether each of the following arguments is valid or not.

(i) The Eiffel Tower is in Australia and Australia is below the equator.
Therefore, the Eiffel Tower is in Australia.

(ii) The Eiffel Tower is in Paris or Paris is in France.
Therefore, the Eiffel Tower is in Paris.

(iii) The Eiffel Tower is in Australia or Australia is below the equator.
Therefore, the Eiffel Tower is in Australia.

(iv) The Eiffel Tower is in Paris and Paris is in France.
Therefore, the Eiffel Tower is in Paris.

(v) The Eiffel Tower is in Australia and Paris is in France.
Therefore, Paris is in France.

(vi) The Eiffel Tower is in Australia or France is in Australia.
Therefore, the Eiffel Tower is in Australia.

(vii) The Eiffel Tower is in Australia or France is in Australia.
Therefore, the Eiffel Tower is in Paris.

(b) For each of the above use your knowledge of English and geography to say whether

(i) the premise, that is the sentence before the 'therefore', is true;

(ii) the conclusion, that is the sentence after the 'therefore', is true.

(c) Can you construct a valid argument with true premise and false conclusion?

2.2.2 Formalizing arguments

The argument forms of the first two Socrates examples can be abstracted to

if p then q

p

therefore

q

and

p

therefore

q

How to represent in logic the natural language form 'if S then T', where S and T are arbitrary statements, has been the subject of heated debate among logicians. It turns out that simple examples of the kind we have here can be written unproblematically as $\mathscr{S} \rightarrow \mathscr{T}$ where $\rightarrow$ is one of the connectives we introduced in the alphabet for propositional logic and $\mathscr{S}$ and $\mathscr{T}$ stand for arbitrary sentences of propositional logic. There are other commonly occurring natural language forms, some of which we have seen above, that correspond to the remaining connectives.

natural language	*propositional logic*
$\mathscr{S}$ and $\mathscr{T}$	$\mathscr{S} \wedge \mathscr{T}$
$\mathscr{S}$ or $\mathscr{T}$	$\mathscr{S} \vee \mathscr{T}$
not $\mathscr{S}$	$\neg \mathscr{S}$
$\mathscr{S}$ if and only if $\mathscr{T}$	$\mathscr{S} \leftrightarrow \mathscr{T}$

We shall see that the correctness of an argument in the propositional calculus is decidable, in other words there are effective procedures for determining whether or not a given argument is correct. One such procedure is the method of **truth-tables** which we now describe.

The Socrates argument can be transcribed into the formal language as

$$((p \rightarrow q) \wedge p) \rightarrow q \qquad \textbf{(2.1)}$$

that is if it is the case both that p is true, and also that if p is true then q is true, then it is the case that q is true.

We have said we are going to assume that every proposition is either true or false and that this is one of the basic assumptions of 'classical' logic – the oldest form of logic and the one that we are studying to start with. Because we are only interested in the form of arguments and the **truth-values** (true or false) of propositions, rather than their more subtle meanings in the everyday world, we use sentence variables to stand for arbitrary interpreted sentences that can take the values true (t) or false (f) alone.

From these we can build up more complicated formulas (with corresponding forms of natural language argument) using the connectives such as those for 'if ... then' and 'and' as shown above.

In elementary algebra, if we know that $a = 2$ and $b = 3$, we can calculate the value of $a + b$ as 5 without knowing or caring whether a stands for 2 apples or 2 houses. Similarly, in classical logic the truth-value of a compound sentence can be calculated solely from the truth-values of its constituent sentence variables, independently of how these were determined or of any other aspects of their meaning in the real world. Logics with this property are called **truth-functional** and their connectives are sometimes called **truth-functors**.

Classical logic has, although as we shall see it does not necessarily need, the five connectives listed above. The rules for using them in truth-value calculations can be conveniently expressed in the form of truth-tables – the same idea as the multiplication tables of elementary arithmetic, but of course the rules are different.

→	t	f
t	t	f
f	t	t

∧	t	f
t	t	f
f	f	f

∨	t	f
t	t	t
f	t	f

↔	t	f
t	t	f
f	f	t

¬	
t	f
f	t

To find, for example, the truth-value of $p \to q$ when the truth-value of p is t and the truth-value of q is f we look along the row marked t for p and down the column marked f for q. The corresponding table entry in the t row and f column is f for $p = \text{t}$ and $q = \text{f}$.

Note that for $\to$ it matters which of its arguments is the row and which the column; we would have got a different answer if we had taken q as the row and p as the column. Some of the truth-functors **commute** and some do not.

With the truth-tables for the connectives the argument 2.1 can be shown correct as follows:

p	q	$p \to q$	$(p \to q) \wedge p$	$((p \to q) \wedge p) \to q$
t	t	t	t	t
t	f	f	f	t
f	t	t	f	t
f	f	t	f	t

What we have done is to build up a truth-table for the sentence that we want to check by using the rules for the connectives to compute the truth-values of all sentences. We see that for any possible combination of truth-values for p and q, that is for all lines in the truth-table, the argument is true. The assignment of truth-values to propositional variables – each line of the truth-table – is called a **valuation**. The sentence 2.1 is true in any valuation and we say therefore that 2.1 is **valid**. Another word often used for sentences that are true in all valuations is **tautology**.

For a two-argument truth-functor there can only be $2^4 = 16$ different truth-tables because for two sentence variables there are only four different combinations of truth-values and each of these can only correspond to one of two truth-values.

In classical logic, that is the logic based on these truth-tables, the connectives we have introduced can be defined in terms of each other. For example $\mathscr{S} \to \mathscr{T}$ is truth-functionally equivalent to $\neg\mathscr{S} \vee \mathscr{T}$. We can show this by the truth-table

$\mathscr{S}$	$\mathscr{T}$	$\mathscr{S} \to \mathscr{T}$	$\neg\mathscr{S} \vee \mathscr{T}$
t	t	t	t
t	f	f	f
f	t	t	t
f	f	t	t

where you see that the column headed $\vee$ is the same as that given above for $\rightarrow$.

It can easily be checked from the truth-table for $\leftrightarrow$ that $\mathcal{S}$ is truth-functionally equivalent to $\mathcal{T}$ if and only if $\mathcal{S} \leftrightarrow \mathcal{T}$ is a tautology, so equivalences are a special kind of tautology in which the principle connective is $\leftrightarrow$. They are useful in the transformation of statements into logically equivalent statements of different, perhaps simpler or more useful, form.

Exercise 2.3

(a) Using the propositional variable r to mean 'it is raining', s to mean 'it is snowing' and f to mean 'it is freezing' translate the following into our logical notation:

(i) It is raining and it is freezing
(ii) It is either raining or snowing, and it is freezing
(iii) It is raining but not snowing
(iv) If it is freezing then it is snowing
(v) It is neither raining nor snowing
(vi) If it is freezing or snowing then it is not raining

(b) How many different n-argument truth-functors are there?

(c) Many other equivalences are known. Here are some. You should check as many as you have the patience for.

(i) $(\mathcal{P} * \mathcal{P}) \leftrightarrow \mathcal{P}$ idempotence.
(ii) $(\mathcal{P} * \mathcal{Q}) \leftrightarrow (\mathcal{Q} * \mathcal{P})$ commutativity.
(iii) $(\mathcal{P} * (\mathcal{Q} * \mathcal{R})) \leftrightarrow ((\mathcal{P} * \mathcal{Q}) * \mathcal{R})$ associativity

where $*$ can be $\wedge$, $\vee$ or $\leftrightarrow$, but not $\rightarrow$.

(iv) Which of (i), (ii) and (iii) does $\rightarrow$ satisfy? Is it for example associative?

Here are some more equivalences.

(v) $(\mathcal{P} \wedge (\mathcal{Q} \vee \mathcal{R})) \leftrightarrow ((\mathcal{P} \wedge \mathcal{Q}) \vee (\mathcal{P} \wedge \mathcal{R}))$.
(vi) $(\mathcal{P} \vee (\mathcal{Q} \wedge \mathcal{R})) \leftrightarrow ((\mathcal{P} \vee \mathcal{Q}) \wedge (\mathcal{P} \vee \mathcal{R}))$.
(vii) $(\mathcal{P} \wedge \neg\mathcal{P}) \leftrightarrow \bot$: a special propositional variable that is false in every valuation (to be pronounced 'absurdity', 'falsity' or 'bottom').
(viii) $(\mathcal{P} \vee \neg\mathcal{P}) \leftrightarrow \top$: a variable that is true in every valuation. This tautology is often called 'excluded middle' or historically, in Latin, *tertium non datur*.
(ix) $\neg\neg\mathcal{P} \leftrightarrow \mathcal{P}$
(x) $\neg(\mathcal{P} \wedge \mathcal{Q}) \leftrightarrow (\neg\mathcal{P} \vee \neg\mathcal{Q})$
(xi) $\neg(\mathcal{P} \vee \mathcal{Q}) \leftrightarrow (\neg\mathcal{P} \wedge \neg\mathcal{Q})$

(d) Use truth-tables to check that the following pairs of sentences are equivalent:

(i) p and $p \vee (p \wedge q)$
(ii) p and $p \wedge (p \vee q)$
(iii) $p \wedge q$ and p

If I now tell you that p means 'everyone loves everyone' and q means 'everyone loves themselves', why is it that given these meanings $p \wedge q$ and p have the same truth-values?

(iv) Use the laws of equivalence above to show that the pairs of sentences in (i) and (ii) are equivalent.

(e) Formalize the arguments from the previous section and use truth-tables to check the assessments of validity that you gave there.

(f) Use truth-tables to determine, for each of the following sentences, whether it is a tautology, a contradiction or contingent, that is neither a tautology nor a contradiction.

(i) $(p \rightarrow (q \rightarrow p))$
(ii) $((p \leftrightarrow \neg q) \vee q)$
(iii) $((p \rightarrow (q \rightarrow r)) \rightarrow ((p \rightarrow q) \rightarrow (p \rightarrow r)))$
(iv) $((p \rightarrow q) \rightarrow (\neg (q \rightarrow p)))$
(v) $((\neg p \rightarrow \neg q) \rightarrow (q \rightarrow p))$
(vi) $\neg (((p \rightarrow q) \rightarrow p) \rightarrow p)$

(g) Write SML and/or Prolog programs to implement a tautology tester for propositional logic sentences using the truth-table method. Try to make your programs as efficient as possible.

2.3 Functional completeness

The logical connectives $\neg$, $\rightarrow$, $\wedge$, $\vee$ and $\leftrightarrow$ are chosen for the fairly natural way that their truth-tables correspond to common usage in argument, but it is not mathematically necessary to have so many.

In fact it can be shown that a single connective is sufficient to represent all classical truth-functors. Such a 'functionally complete' connective is the Sheffer stroke, written |, whose truth table is the same as that for $\neg(\mathscr{S} \wedge \mathscr{T})$.

Another connective with this property of being functionally complete is 'NOR', which is equivalent to $\neg(\mathscr{S} \vee \mathscr{T})$. This has important consequences in hardware design. A digital circuit can be built to realize any possible truth-functor from NOR gates alone (circuits which realize NOR).

Exercise 2.4

(a) Show that

(i) $\neg \mathscr{S} \leftrightarrow \mathscr{S} \mid \mathscr{S}$
(ii) $\mathscr{S} \wedge \mathscr{T} \leftrightarrow (\mathscr{S} \mid \mathscr{T}) \mid (\mathscr{S} \mid \mathscr{T})$

(b) How can $\mathscr{S} \vee \mathscr{T}$ and $\mathscr{S} \rightarrow \mathscr{T}$ be written in terms of | alone?

2.4 Consistency, inconsistency, entailment

Definition 2.2: A **valuation** is an assignment of truth-values to the propositional variables.

Thus a valuation is a function $v: P \rightarrow \{t, f\}$, where P is the set of propositional variables.

Definition 2.3: A sentence is **valid** iff it is true (evaluates to t) for all valuations. A valid sentence is also called a **tautology**.

Definition 2.4: Suppose you have a set

$$G = \{\mathscr{S}_1, \ldots, \mathscr{S}_n\}$$

of sentences. In some valuations (lines of the truth-table) some of the $\mathscr{S}_i$ will be true and some false. If there is at least one line in which all the $\mathscr{S}_i \in G$ are true then we say that G is semantically **consistent**. If no valuation makes all the $\mathscr{S}_i$ true then G is **inconsistent**.

An example of inconsistency is the set $\{\mathscr{S}, \neg\mathscr{S}\}$.

Definition 2.5: A set of sentences G **semantically entails** a sentence $\mathscr{T}$, written as $G \models \mathscr{T}$, iff there is no valuation that makes all of the sentences in G true and makes $\mathscr{T}$ false, that is assuming the truth of all $\mathscr{S}_i \in G$ has the consequence that $\mathscr{T}$ is true.

If we now adopt the convention that we write $\models \mathscr{T}$ in the case where G is the empty set then we can use $\models \mathscr{T}$ as a neat notation for saying that $\mathscr{T}$ is a tautology because if G is empty then the definition of semantic entailment says that there is no valuation which makes $\mathscr{T}$ false, so $\mathscr{T}$ is true in all valuations. Also, if $\mathscr{T}$ is omitted, then the definition can be read as saying that there is no valuation which makes all of the sentences in G true, so we can write $G\models$ for G is inconsistent.

With this convention, $\models$ (which of course is a symbol in the observer's language for a relation between sentences and sets of sentences in the object language) behaves rather like the $=$ sign in algebraic equations. We have $G \models \mathscr{T}$ iff $G, \neg\mathscr{T} \models$ where $G, \neg\mathscr{T}$ is a convenient abbreviation for $G \cup \{\neg\mathscr{T}\}$.

We can think of this as formalizing the idea that, given the **assumptions** in G, then the **conclusion** $\mathscr{T}$ is true, or $\mathscr{T}$ follows from the assumptions.

As examples, we look at the sets $\{p, \neg p\}$ and $\{(p \wedge q) \rightarrow q, \neg p \vee \neg q, p\}$ and try to decide on their consistency. To do this we write out the truth-tables, which enumerate all possible valuations for the propositional variables p and q, to see whether or not there is at least one row in the truth-tables which has t in each place, that is to see whether or not all the sentences can simultaneously be true.

In the first case we get

p	$\neg p$
t	f
f	t

and here we can clearly see that no row has t everywhere, so the sentences in $\{p, \neg p\}$ are inconsistent. Next we get

p	q	$(p \wedge q)$	$\neg p$	$\neg q$	$(p \wedge q) \rightarrow q$	$(\neg p \vee \neg q)$
t	t	t	f	f	t	f
t	f	f	f	t	t	t
f	t	f	t	f	t	t
f	f	f	t	t	t	t

and we can see that the set $\{(p \wedge q) \rightarrow q, \neg p \vee \neg q, p\}$ is consistent because the second line of the truth-table shows that all three sentences in the set have value t.

We next give some examples of deciding the correctness of questions of semantic entailment. Remember that $G \vDash \mathscr{S}$ iff there is no valuation which, while making all of the sentences in G true, makes $\mathscr{S}$ false. This, again, can be decided by writing out truth-tables so as to enumerate all possible valuations.

Consider the examples $\{p\} \vDash p$, $\{p \rightarrow q, p\} \vDash q$ and $\{p, \neg p\} \vDash q$. The first is trivial since the truth-table is just

p
t
f

and here everywhere p has value t then p has value t. The second case gives the truth-table

p	q	$p \rightarrow q$
t	t	t
t	f	f
f	t	t
f	f	t

and we can see that in the single case, in the first line, where the assumptions are both true then the conclusion is too, so the entailment is valid. Finally we have

p	q	$\neg p$
t	t	f
t	f	f
f	t	t
f	f	t

and we can see that nowhere do the assumptions both have value t and this means that the conclusion never has the value f when all the assumptions have the value t. Hence, the entailment is valid.

Note that this would always be the outcome if the assumptions are $\{p, \neg p\}$ since, as we saw above, this set is inconsistent, so it is never the case that both sentences have value t simultaneously. Any entailment therefore, no matter what the conclusion, would be valid with these assumptions. We can sum this up by saying that 'anything follows from inconsistent assumptions'.

Exercise 2.5

(a) Work out which of the following sets of sentences are semantically consistent and which are semantically inconsistent:

(i) $\{p \to q, \neg q\}$
(ii) $\{p \to q, \neg q \vee r, p \wedge \neg r\}$
(iii) $\{(p \vee q) \to r, \neg((\neg p \wedge \neg q) \vee r)\}$

(b) Work out which of the following semantic entailments are correct and which are not.

(i) $\{p \to q\} \models (\neg q \to \neg p)$
(ii) $\{p \to q\} \models q$
(iii) $\{(p \vee q) \to r, \neg r\} \models \neg p$
(iv) $\models ((p \to q) \to p) \to p$

(c) Rewrite $(a \vee b) \wedge (a \vee c)$ to a logically equivalent sentence containing no connectives other than $\neg$ and $\to$.

(d) If $\mathscr{A}$ and $\mathscr{B}$ are any sentences and T is a set of sentences then show that $T \cup \{\mathscr{A}\} \models \mathscr{B}$ if and only if $T \models \mathscr{A} \to \mathscr{B}$.

(e) Here are some facts about entailment; you should prove that they follow from the definition. $\mathscr{R}$, $\mathscr{S}$ and $\mathscr{T}$ are any sentences and G and D are any sets of sentences. Recall that G, D means $G \cup D$ and $G, \mathscr{T}$ means $G \cup \{\mathscr{T}\}$.

(i) $G, \mathscr{T} \models \mathscr{T}$
(ii) if $G \models \mathscr{T}$ then $G, D \models \mathscr{T}$ (monotonicity)
(iii) if $G \models \mathscr{T}$ and $G, \mathscr{T} \models \mathscr{S}$ then $G \models \mathscr{S}$ (called the cut theorem)
(iv) if $\mathscr{T} \models \mathscr{S}$ and $\mathscr{S} \models \mathscr{R}$ then $\mathscr{T} \models \mathscr{R}$ (transitivity)
(v) $G \models \neg \mathscr{T}$ iff $G, \mathscr{T} \models$
(vi) $G \models \mathscr{T}$ and $G \models \mathscr{S}$ iff $G \models (\mathscr{T} \wedge \mathscr{S})$

(vii) $G, \mathcal{T} \models \mathcal{R}$ and $G, \mathcal{S} \models \mathcal{R}$ iff $G, (\mathcal{T} \vee \mathcal{S}) \models \mathcal{R}$
(viii) $G, \mathcal{T} \models \mathcal{S}$ iff $G \models \mathcal{T} \rightarrow \mathcal{S}$
(ix) $G, \mathcal{T} \models \mathcal{S}$ and $G, \mathcal{S} \models \mathcal{T}$ iff $G \models \mathcal{T} \leftrightarrow \mathcal{S}$

(f) Use the properties of $\models$ listed in question (e) to show, without using truth-tables, that:
(i) $\models p \rightarrow (q \rightarrow p)$
(ii) $\models (p \rightarrow q) \rightarrow ((q \rightarrow r) \rightarrow (p \rightarrow r))$

(g) Three people A, B and C are apprehended on suspicion of cruelty to mice in the computer laboratory.

A says: 'B did it; C is innocent'

B says: 'If A is guilty then so is C'

C says: 'I didn't do it; one of the others did'

(i) Are the statements consistent?
(ii) Assuming that everyone is innocent of the dastardly deed, who told lies?
(iii) Assuming that everyone's statement is true, who is innocent and who is guilty?

Hint: let I_A stand for 'A is innocent', I_B for 'B is innocent' and I_C for 'C is innocent'. Then A's statement can be written as $\neg I_B \wedge I_C$, which we call S_A, and so on.

For each valuation of I_A, I_B and I_C determine the truth-values of S_A, S_B and S_C. All these answers can be determined from the truth-table.

(h) The cut theorem, which is one of the facts about entailment that you were invited to prove above, says:

if $G \models \mathcal{T}$ and $G, \mathcal{T} \models \mathcal{S}$ then $G \models \mathcal{S}$

where $\mathcal{T}$ and $\mathcal{S}$ are sentences and G is a set of sentences. Give a counterexample, that is examples of G, $\mathcal{T}$ and $\mathcal{S}$, which shows that if $G, \mathcal{T} \models \mathcal{S}$ then $G \models \mathcal{S}$ does not always hold.

2.5 Formal propositional calculus

Up to now we have been working with **informal** systems. We now re-express all of the above in more formal terms and then go on to introduce our first formal system. That will mean that the person using the system no longer has to know the **meaning** of the symbols in the language or the meaning behind any rule – the only thing communicated is the **form** of the expressions and rules, and the relations between them. To use a formal system correctly, and hence to make arguments correctly, we need only follow the rules correctly. If this can be done then, because the rules are completely specified, it follows that a **program** can be written which can construct correct proofs.

To show that an argument is correct, in other words to assess its validity or otherwise, we first need to give a formal interpretation for the symbols of the language. This is where the notion of valuation, as already introduced, will be used. But then a further development will be to construct a valid argument step by step without knowing the meaning of the expressions and rules. This is where we will turn to the formal deductive system.

2.5.1 The language of propositions

We will first elaborate some of the definitions which were given fairly informally in previous sections.

Definition 2.6: A propositional language is based on two components, an **alphabet** and a **grammar**.

(a) The **alphabet** consists of three sets:

- (i) a set of connective symbols $\{\neg, \rightarrow, \wedge, \vee, \leftrightarrow\}$,
- (ii) a set of punctuation symbols $\{(,)\}$,
- (iii) a set of propositional variables P.

(b) The **grammar**, which defines what counts as being a **sentence based on P**, is given by

- (i) each of the elements of P is a sentence based on P,
- (ii) if $\mathscr{S}$ and $\mathscr{T}$ are sentences based on P then so are $(\neg\mathscr{S})$, $(\mathscr{S} \wedge \mathscr{T})$, $(\mathscr{S} \vee \mathscr{T})$, $(\mathscr{S} \rightarrow \mathscr{T})$ and $(\mathscr{S} \leftrightarrow \mathscr{T})$,
- (iii) nothing else is a sentence based on P.

Definition 2.7: The set of sentences based on P is called the **propositional language** based on P, which we write as $L(P)$. We often refer to the elements of P as **atomic** sentences.

We have already seen one particular way, based on truth-tables, of giving a meaning to the symbols in a language. We are now going to be more abstract so, instead of tying ourselves to one particular way of giving meaning, we simply present a general definition of what it is to give meaning to a language. The truth-tables can be seen as a particular example of this.

Recall that a valuation is a function v from a set of propositional variables to the set $\{t, f\}$. If the domain of v is P then we say that v is a ***P*-valuation**.

We can now extend v into a new function which gives meaning not only to the members of P, as v does, but to any member of $L(P)$. This extended function, $\text{truthvalue}_v: L(P) \rightarrow \{t, f\}$, is defined below by cases based on the structure of members of $L(P)$. By convention, we drop the v subscript where it is clear which valuation we are basing our definition on.

Definition 2.8:

(i) $\text{truthvalue}(\mathscr{S}) = v(\mathscr{S})$ if $\mathscr{S}$ is atomic.

(ii) truthvalue($\neg\mathscr{S}$) = t if truthvalue($\mathscr{S}$) = f and truthvalue($\neg\mathscr{S}$) = f if truthvalue($\mathscr{S}$) = t.

(iii) truthvalue($\mathscr{S} \wedge \mathscr{T}$) = t if truthvalue($\mathscr{S}$) = t and truthvalue($\mathscr{T}$) = t, and truthvalue($\mathscr{S} \wedge \mathscr{T}$) = f otherwise.

(iv) truthvalue($\mathscr{S} \vee \mathscr{T}$) = t if truthvalue($\mathscr{S}$) = t or truthvalue($\mathscr{T}$) = t, or both, and truthvalue($\mathscr{S} \vee \mathscr{T}$) = f otherwise.

(v) truthvalue($\mathscr{S} \rightarrow \mathscr{T}$) = f if truthvalue($\mathscr{S}$) = t and truthvalue($\mathscr{T}$) = f, and truthvalue($\mathscr{S} \rightarrow \mathscr{T}$) = t otherwise.

(vi) truthvalue($\mathscr{S} \leftrightarrow \mathscr{T}$) = t if truthvalue($\mathscr{S}$) = truthvalue($\mathscr{T}$) and truthvalue($\mathscr{S} \leftrightarrow \mathscr{T}$) = f otherwise.

As an example, consider the sentence $(\neg(p \vee q) \rightarrow (r \wedge p))$ in a context where we have a valuation v such that $v(p)$ = t, $v(q)$ = f and $v(r)$ = t. We have

truthvalue($(\neg(p \vee q) \rightarrow (r \wedge p))$) depends on
  truthvalue($\neg(p \vee q)$) which depends on
    truthvalue($(p \vee q)$) which depends on
      truthvalue(p) = $v(p)$ = t
        or truthvalue(q) = $v(q)$ = f
      so truthvalue($(p \vee q)$) = t
    so truthvalue($\neg(p \vee q)$) = f
  and truthvalue($(r \wedge p)$) which depends on
    truthvalue(r) = $v(r)$ = t
      and truthvalue(p) = $v(p)$ = t
    so truthvalue($(r \wedge p)$) = t
  so truthvalue($(\neg(p \vee q) \rightarrow (r \wedge p))$) = t.

Now this computation of the truth-value of a sentence, in the context of a certain valuation, clearly follows the same pattern as the computation which sets out to decide whether or not a sequence of symbols from the alphabet is a member of the language.

If we recall the definition of sentencehood above, and use it on the same sentence $(\neg(p \vee q) \rightarrow (r \wedge p))$ as above, we get

decide($(\neg(p \vee q) \rightarrow (r \wedge p))$) depends on
  decide($\neg(p \vee q)$) which depends on
    decide($(p \vee q)$) which depends on
      decide(p) = true, since $p \in P$
        and decide(q) = true, since $q \in P$
      so decide($(p \vee q)$) = true
    so decide($\neg(p \vee q)$) = true
  and decide($(r \wedge p)$) which depends on
    decide(r) = true, since $r \in P$

and decide(p) = true, since $p \in P$
so decide($(r \wedge p)$) = true
so decide($(\neg(p \vee q) \rightarrow (r \wedge p))$) = true.

This similarity of computations should come as no surprise, since their patterns are governed by the grammar, which is the same in each case.

Exercise 2.6

(a) How many sentences are there for the alphabet which has $P = \{q\}$?

(b) How many sentences are there in the language which has $P = \{p_1, p_2, \ldots\}$, that is a countably infinite set?

(c) For each of the following sentences, and given the valuation v where $v(p) = \text{t}$, $v(q) = \text{f}$ and $v(r) = \text{t}$, find the truth-values of
 (i) $(p \vee \neg p)$
 (ii) $(p \wedge \neg p)$
 (iii) $(p \leftrightarrow \neg\neg p)$
 (iv) $((p \rightarrow q) \rightarrow (p \vee q))$
 (v) $(((p \vee q) \wedge r) \rightarrow (q \leftrightarrow p))$

(d) Write an SML function which implements truthvalue. Make it look as much like the SML function for decide as you can.

2.5.2 More definitions

Definition 2.9: If v is a P-valuation we say that v **satisfies** a sentence $\mathscr{S}$ of $L(P)$ if truthvalue$_v(\mathscr{S}) = \text{t}$, which is also written $\models_v \mathscr{S}$. A P-valuation v satisfies a set of sentences $\{\mathscr{S}_1, \ldots, \mathscr{S}_n\}$ if truthvalue$_v(\mathscr{S}_i) = \text{t}$ for all i between 1 and n. Note that this is the same as saying truthvalue$_v(\mathscr{S}_1 \wedge \ldots \wedge \mathscr{S}_n) = \text{t}$.

If G and D are two sets of sentences then v satisfies $G \cup D$ if and only if v satisfies both G and D. To make this work in general we say that any P-valuation satisfies the empty set.

Definition 2.10: A P-valuation v is a **model** for a set of sentences iff v satisfies the set of sentences.

Definition 2.11: If $\mathscr{S} \in L(P)$ then $\mathscr{S}$ is a **tautology** if and only if truthvalue$_v(\mathscr{S}) = \text{t}$ for all v. Equivalently, $\mathscr{S}$ is a tautology if and only if $\models_v \mathscr{S}$ for all v. As before we write this simply as $\models \mathscr{S}$.

2.5.3 Formal deductive systems

Definition 2.12: A formal system has the following components:

(a) an **alphabet** of symbols,

(b) rules for building sentences, the **grammar**,

(c) a set of sentences, the **axioms**,

(d) a finite set of rules for generating new sentences from old, the **rules of deduction**.

The main characteristic of a formal system is that the behaviour and properties of the symbols in the alphabet are given entirely by the rules of deduction. The symbols themselves have no meaning so the notion of valuation, which we were using above, does not arise. However, as we shall see later, a desirable property of a formal deductive system is that it bears a particular relationship to any valuation.

2.5.4 Formal deductive system for propositional calculus

The alphabet and grammar components will be just as they were in the previous sections. The two new components are the axioms and the rules of inference. The axiom system that we give here is not the only one possible. A set of axiom schemas for propositional calculus was first given by Frege; it contained six sentences. The set that we have chosen was given, and shown to do the same job as Frege's, by Lukasiewicz. Yet another, this time with four axiom schemas, was given by Hilbert and Ackermann. In fact, the one we give is not the smallest possible: Nicod used the Sheffer stroke to give a set consisting of just one schema, though instead of modus ponens he used a slightly more complicated variant of it.

(a) the alphabet;

(b) the grammar;

(c) the axioms: if $\mathcal{S}$, $\mathcal{T}$ and $\mathcal{R}$ are any sentences then

(A1) $(\mathcal{S} \rightarrow (\mathcal{T} \rightarrow \mathcal{S}))$

(A2) $((\mathcal{S} \rightarrow (\mathcal{T} \rightarrow \mathcal{R})) \rightarrow ((\mathcal{S} \rightarrow \mathcal{T}) \rightarrow (\mathcal{S} \rightarrow \mathcal{R})))$

(A3) $(((\neg \mathcal{S}) \rightarrow (\neg \mathcal{T})) \rightarrow (\mathcal{T} \rightarrow \mathcal{S}))$

are **axiom schemas**, which generate the axioms in the same way as the clauses of the grammar generate the sentences;

(d) rule of deduction, which for historical reasons is called **modus ponens**, often abbreviated to MP: from $\mathcal{S}$ and $(\mathcal{S} \rightarrow \mathcal{T})$, for $\mathcal{S}$ and $\mathcal{T}$ any sentences, we obtain $\mathcal{T}$ as a **direct consequence**.

So we actually have infinitely many axioms since $\mathcal{S}$, $\mathcal{T}$ and $\mathcal{R}$ can be any sentences and the number of sentences is infinite. Clearly, though, there are more sentences than there are axioms. For example $(p \rightarrow p)$ is certainly a sentence but it is not an axiom. Just as clearly, all the axioms are sentences. These two facts together mean that the set of axioms is a proper subset of the set of sentences, yet both are infinitely large. However, this is not such a strange situation; the natural numbers form an infinite set and the even natural numbers form another infinite set which is a proper subset of the natural numbers. The similarity can be seen if we give their set definitions:

$$\mathrm{Ax}(P) = \{\mathscr{S} \mid \mathscr{S} \in L(P) \text{ and axiom}(\mathscr{S})\}$$
$$\mathrm{Evens} = \{n \mid n \in N \text{ and even}(n)\}$$

Here axiom($\mathscr{S}$) is true whenever $\mathscr{S}$ is an axiom, and even(n) is true whenever there is a natural number k such that $n = 2k$. We can see that even is decidable, but can we define axiom so that it is decidable too? Well, once again we can look to the structure of the definition of the axiom schemas for an answer. The definition above can be recast so that axiom is

axiom P sentence = axiom1 P sentence or axiom2 P sentence or axiom3 P sentence

where

axiom1 P $(\mathscr{S} \to (\mathscr{T} \to \mathscr{R}))$ = true if decide P $\mathscr{S}$ and decide P $\mathscr{T}$ and $\mathscr{S} = \mathscr{R}$
axiom2 P $((\mathscr{S} \to (\mathscr{T} \to \mathscr{R})) \to ((\mathscr{S}' \to \mathscr{T}') \to (\mathscr{S}'' \to \mathscr{R}')))$ = true if decide P $\mathscr{S}$ and decide P $\mathscr{T}$ and decide P $\mathscr{R}$ and $\mathscr{S} = \mathscr{S}'$ and $\mathscr{R} = \mathscr{R}'$ and $\mathscr{T} = \mathscr{T}'$ and $\mathscr{S} = \mathscr{S}''$
axiom3 P $(((\neg\mathscr{S}) \to (\neg\mathscr{T})) \to (\mathscr{T}' \to \mathscr{S}'))$ = true if decide P $\mathscr{S}$ and decide P $\mathscr{T}$ and $\mathscr{S} = \mathscr{S}'$ and $\mathscr{T} = \mathscr{T}'$

and axiom is false for all other sentences (P is the set of propositional variables in the alphabet). Finally, since decide is decidable it follows that axiom is too.

The ML program for all this is:

```
fun axiom1 P (Imp(S,Imp(T,S'))) = decide P S andalso decide P T andalso (S = S')
  | axiom1 P s = false;

fun axiom2 P (Imp(Imp(S,Imp(T,R)),Imp(Imp(S',T'),Imp(S'',R')))) =
            decide P S andalso decide P T andalso decide P R andalso
                (S = S') andalso (T = T') andalso (R = R') andalso (S = S'')
  | axiom2 P s = false;

fun axiom3 P (Imp(Imp(Not S,Not T),Imp(T',S'))) = decide P S andalso
                                decide P T andalso (S = S') andalso (T = T')
  | axiom3 P s = false;

fun axiom P s = axiom1 P s orelse axiom2 P s orelse axiom3 P s;
```

There is another way of deciding whether or not a sentence is an axiom and that is by introducing the idea of instances. First, we need some definitions and notation.

Let P and Q be two sets of propositional variables. Then these give rise to two propositional languages $L(P)$ and $L(Q)$.

Let vars_P be a function which takes an element of a propositional language as argument and returns as result the set of all those propositional variables that appear in the sentence. So, for example, if $P = \{p_1, p_2\}$ then $\text{vars}_P(p_1 \rightarrow (p_2 \rightarrow p_2)) = \{p_1, p_2\}$.

We can give a method for calculating vars_P which, again, is based on the structure of sentences:

$$\begin{aligned}\text{vars}_P(\mathcal{S}) = \{\mathcal{S}\} &\text{ if } \mathcal{S} \in P \\ \text{or} \quad &\text{vars}_P(\mathcal{T}) \text{ if } \mathcal{S} = \neg\mathcal{T} \\ \text{or} \quad &\text{vars}_P(\mathcal{T}) \cup \text{vars}_P(\mathcal{R}) \text{ if } \mathcal{S} = \mathcal{T} \wedge \mathcal{R} \text{ or } \mathcal{T} \rightarrow \mathcal{R} \text{ or } \\ &\mathcal{T} \vee \mathcal{R} \text{ or } \mathcal{T} \leftrightarrow \mathcal{R}\end{aligned}$$

Now, if $\text{vars}_P(\mathcal{S}) = \{p_1, \ldots, p_k\}$ then $\mathcal{S}[b_1/p_1, \ldots, b_k/p_k]$, where $b_i \in L(Q)$, is the sentence of $L(Q)$ that results from simultaneously substituting b_i for p_i for all i from 1 to k. We say that $\mathcal{S}[b_1/p_1, \ldots, b_k/p_k]$ is a ***Q*-instance** of $\mathcal{S}$, which is a sentence based on P (as usual, when there is sufficient context we omit the reference to the sets P and Q).

Now, to get back to our axioms, let $\mathcal{P} = \{\mathcal{S}, \mathcal{T}, \mathcal{R}\}$. Consider the axiom schemas above as the (only) three sentences of a language based on $\mathcal{P}$, which we call $l(\mathcal{P})$. Then the axioms as defined above are simply all the P-instances of all the elements of $l(\mathcal{P})$. That is

$$\begin{aligned}\text{Ax}(P) = \{\mathcal{S}[\mathcal{S}_1/p_1, \ldots, \mathcal{S}_n/p_n] \in L(P) \mid \mathcal{S} \in l(\mathcal{P}) \\ \text{and } \text{vars}_{\mathcal{P}}(S) = \{\mathcal{S}_1, \ldots, \mathcal{S}_n\} \\ \text{and } p_1, \ldots, p_n \in L(P)\}\end{aligned}$$

The idea of an instance is probably quite an intuitive one. In fact, we have been using it implicitly all along by using script letters to range over all sentences. What we were doing, we can now see, was to express results, definitions and questions in a language based on an alphabet of script letters and to assume that it was clear how to apply these general results in particular cases, that is how the results, definitions and questions related to all instances over an appropriate alphabet.

2.5.5 Proofs

Definition 2.13: A **proof** in a formal deductive system (FDS) is a sequence of sentences $\mathcal{S}_1, \ldots, \mathcal{S}_n$ such that, for all $i \leqslant n$, either $\mathcal{S}_i$ is an axiom or there are two members $\mathcal{S}_j$, $\mathcal{S}_k$ of the sequence, with $j, k < i$, which have $\mathscr{S}_i$ as a direct consequence by modus ponens (MP). $\mathscr{S}_n$ is then a **theorem** of the FDS, and the sequence $\mathscr{S}_1, \ldots, \mathscr{S}_n$ is a proof of $\mathscr{S}_n$.

Note that, in such a proof, any of the $\mathscr{S}_i$ are theorems since we can truncate the proof at $\mathscr{S}_i$, giving a proof of $\mathscr{S}_i$.

Definition 2.14: If G is a set of sentences then a sequence $\mathscr{S}_1, \ldots, \mathscr{S}_n$ is a **deduction from G** if, for each $i \leqslant n$

(a) $\mathscr{S}_i$ is an axiom, or

(b) $\mathscr{S}_i \in G$, or

(c) $\mathscr{S}_i$ follows from two earlier members of the sequence as a direct consequence by MP.

We can think of a deduction from G as being like a proof with the elements of G as temporary axioms.

Definition 2.15: If G is a set of sentences and $\mathscr{S}_1, \ldots, \mathscr{S}_n$ is a deduction from G then $\mathscr{S}_n$ is **deducible from G**, and this is written $G \vdash \mathscr{S}_n$.

'$\mathscr{S}$ is a theorem' can therefore be written as $\vdash \mathscr{S}$, that is $\mathscr{S}$ is deducible from the empty set. And, as should be clear from the definition by now, we could write a program which takes any sequence of sentences and decides whether or not the sequence is a proof. That is, proofhood is also decidable.

EXAMPLE 2.1

Here is an example of a proof of the sentence $(p \rightarrow p)$. Note that the numbers on the left and the comments on the right are not part of the formal proof, they are annotations to help the reader see that the steps are justified.

(1)	$((p \rightarrow ((p \rightarrow p) \rightarrow p)) \rightarrow ((p \rightarrow (p \rightarrow p)) \rightarrow (p \rightarrow p)))$	instance of A2
(2)	$(p \rightarrow ((p \rightarrow p) \rightarrow p))$	instance of A1
(3)	$((p \rightarrow (p \rightarrow p)) \rightarrow (p \rightarrow p))$	MP with (1) and (2)
(4)	$(p \rightarrow (p \rightarrow p))$	instance of A1
(5)	$(p \rightarrow p)$	MP with (3) and (4)

Even this simple example shows that proofs in this system are difficult. This is mainly because very long sentences are generated and it is usually hard to see how to proceed correctly or what are the appropriate axiom instances to choose. Later we shall see some far better syntactic proof methods, but we introduce the system here because it is easy to prove things about it (if not in it).

EXAMPLE 2.2

Now we have an example of a proof that uses assumptions. Recall from above that the assumptions can be used in the same way as the axioms.

We show that $\{p, (q \rightarrow (p \rightarrow r))\} \vdash (q \rightarrow r)$.

(1)	p	assumption
(2)	$(q \rightarrow (p \rightarrow r))$	assumption
(3)	$(p \rightarrow (q \rightarrow p))$	instance of A1
(4)	$(q \rightarrow p)$	MP with (1) and (3)
(5)	$((q \rightarrow (p \rightarrow r)) \rightarrow ((q \rightarrow p) \rightarrow (q \rightarrow r)))$	instance of A2
(6)	$((q \rightarrow p) \rightarrow (q \rightarrow r))$	MP with (2) and (5)
(7)	$(q \rightarrow r)$	MP with (4) and (6)

Exercise 2.7

(a) For each stage of the two proofs above, write down the substitutions used in the relevant axiom schemas to obtain the sentences in the proof.

(b) Construct proofs of the following:

(i) $\vdash (a \rightarrow (a \rightarrow a))$
(ii) $\{a\} \vdash a \rightarrow a$
(iii) $\{a\} \vdash a \rightarrow a$, but make it different from the proof you gave in (ii)
(iv) $\{a, a \rightarrow b\} \vdash b$
(v) $\{a\} \vdash b \rightarrow a$
(vi) $\{f \rightarrow g, g \rightarrow h\} \vdash f \rightarrow h$
(vii) $\{f \rightarrow (g \rightarrow h), g\} \vdash f \rightarrow h$
(viii) $\vdash \neg\neg f \rightarrow f$
(ix) $\vdash f \rightarrow \neg\neg f$

(c) Write an SML or Prolog program which takes a proof and says whether it is valid or not. You should obviously use the programs for `axiom` given above, amongst other things.

(d) Write an SML or Prolog program which implements $vars_p$ as above.

(e) Write an SML or Prolog program which implements a function `instance_of` which takes a sentence S, a propositional variable p and another propositional variable q and substitutes all occurrences of p in S by q. Clearly this function will be based on the structure of sentences. Try to extend the function so that it takes a list of propositional variable pairs as described in the text above.

2.6 Soundness and completeness for propositional calculus

Now that we have a system in which we can build proofs, we would like to be sure that the sentences we can find proofs for, the theorems, are indeed the tautologies that we previously characterized as valid. So, we have to show that the following is true:

> For any sentence $\mathscr{S}$, if $\mathscr{S}$ is a theorem then $\mathscr{S}$ is a tautology and if $\mathscr{S}$ is a tautology then $\mathscr{S}$ is a theorem.

This will allow us to claim that our formal system adequately characterizes our intuitive ideas about valid arguments, so we say that the above statement expresses the **adequacy theorem** for the formal system. The first part expresses the **soundness** of our formal system. It says that if we show that $\mathscr{S}$ is a theorem then it is a tautology. In other words we only ever produce valid sentences as theorems. Conversely, the second part expresses **completeness** and says that any $\mathscr{S}$ which is a tautology will always be provable in the formal system.

2.6.1 The deduction theorem

This section introduces and proves a theorem that turns out to be very useful for both drastically shortening proofs within the formal system and also for easing the presentation of the completeness proof, which we do later.

> ***Theorem***
>
> If $G, \mathscr{S} \vdash \mathscr{T}$ then $G \vdash \mathscr{S} \to \mathscr{T}$, where $\mathscr{S}$ and $\mathscr{T}$ are any sentences and G is any set of sentences.

Proof Let $\mathscr{T}_1, \ldots, \mathscr{T}_n$ be a proof of $\mathscr{T}$ from assumptions $G \cup \{\mathscr{S}\}$, so $\mathscr{T}_n$ is $\mathscr{T}$. We do induction on n.

If $n = 1$ then $\mathscr{T}$ is in G or $\mathscr{T}$ is an instance of an axiom schema or $\mathscr{T}$ is $\mathscr{S}$. In the first two cases $G \vdash \mathscr{T}$ so since $\vdash \mathscr{T} \to (\mathscr{S} \to \mathscr{T})$ we have $G \vdash \mathscr{S} \to \mathscr{T}$. In the final case since $\vdash \mathscr{S} \to \mathscr{T}$ we have $G \vdash \mathscr{S} \to \mathscr{T}$.

Now assume that for any k, $1 \leqslant k < n$, the result holds. There are four possibilities: $\mathscr{T}$ is in G, or $\mathscr{T}$ is an instance of an axiom schema, or $\mathscr{T}$ is $\mathscr{S}$, or $\mathscr{T}$ follows as a direct consequence by modus ponens from $\mathscr{T}_i$ and $\mathscr{T}_j$, $1 \leqslant i, j < n$. In the first three cases we can argue as we did in the case where $n = 1$.

In the fourth case $\mathscr{T}_j$, say, has the form $\mathscr{T}_i \to \mathscr{T}_n$. By the assumption we have $G \vdash \mathscr{S} \to \mathscr{T}_i$ and $G \vdash \mathscr{S} \to (\mathscr{T}_i \to \mathscr{T}_n)$. But an instance of axiom schema A2 is $\vdash (\mathscr{S} \to (\mathscr{T}_i \to \mathscr{T}_n)) \to ((\mathscr{S} \to \mathscr{T}_i) \to (\mathscr{S} \to \mathscr{T}))$, so we have by modus ponens, using $G \vdash \mathscr{S} \to (\mathscr{T}_i \to \mathscr{T}_n)$, that $G \vdash (\mathscr{S} \to \mathscr{T}_i) \to (\mathscr{S} \to \mathscr{T})$ and then by MP again, this time with $G \vdash \mathscr{S} \to \mathscr{T}_i$, that $G \vdash \mathscr{S} \to \mathscr{T}$. So, by induction on n, the theorem holds. ■

2.6.2 The adequacy theorem

There are two parts to this proof since we have to show both if $\vdash \mathscr{S}$ then $\models \mathscr{S}$ and if $\models \mathscr{S}$ then $\vdash \mathscr{S}$. We also use, and prove, a couple of lemmas.

Lemma 2.1

If $\mathscr{S}$ and $(\mathscr{S} \rightarrow \mathscr{T})$ are both tautologies then $\mathscr{T}$ is a tautology.

Proof Assume that $\mathscr{T}$ is not a tautology. We will now deduce a contradiction, which will mean that our initial assumption is false, so $\mathscr{T}$ must be a tautology. (This method of argument is known as 'reductio ad absurdum'.)

If $\mathscr{T}$ is not a tautology, but both $\mathscr{S}$ and $(\mathscr{S} \rightarrow \mathscr{T})$ are, then there must be a valuation such that $v(\mathscr{S}) = \mathrm{t}$, $v((\mathscr{S} \rightarrow \mathscr{T})) = \mathrm{t}$ and $v(\mathscr{T}) = \mathrm{f}$.

If $v((\mathscr{S} \rightarrow \mathscr{T})) = \mathrm{t}$, then $v(\mathscr{S}) = \mathrm{f}$ or $v(\mathscr{T}) = \mathrm{t}$ or both. If $v(\mathscr{S}) = \mathrm{f}$ then this contradicts $v(\mathscr{S}) = \mathrm{t}$. If $v(\mathscr{T}) = \mathrm{t}$ then this contradicts $v(\mathscr{T}) = \mathrm{f}$.

Either way we have a contradiction so $\mathscr{T}$ must be a tautology. ■

With this, we can go on to give the soundness theorem.

Theorem

If $\vdash \mathscr{S}$ then $\models \mathscr{S}$, where $\mathscr{S}$ is any sentence.

Proof If $\mathscr{S}$ is a theorem then there is a sequence $\mathscr{S}_1, \ldots, \mathscr{S}_n$ with $\mathscr{S}_n = \mathscr{S}$ such that the sequence is a proof of $\mathscr{S}$. If $\mathscr{S}$ is an axiom then there is nothing to prove since it can be shown by straightforward calculation of their truth-values that all our axioms are tautologies. If $\mathscr{S}$ is not an axiom then we proceed by induction on the length of the proof of $\mathscr{S}$.

If $n = 1$ then $\mathscr{S}$ is an axiom (by the definition of proof) and so $\mathscr{S}$ is a tautology. If $n > 1$ then we proceed by assuming that all theorems with proofs of length less than n are tautologies.

Now, either $\mathscr{S}$ is an axiom or the proof of $\mathscr{S}$ contains two sentences $\mathscr{S}_i$ and $\mathscr{S}_j$, with $i, j < n$, such that $\mathscr{S}_j$ (say) is $(\mathscr{S}_i \rightarrow \mathscr{S})$. $\mathscr{S}_i$ and $\mathscr{S}_j$ are tautologies by hypothesis (since they have proofs of length less than n) and so, by Lemma 2.1, $\mathscr{S}$ is a tautology too. So the induction step is complete and the soundness theorem is proved. ■

Before we go on to the completeness theorem we need the other lemma.

Lemma 2.2

Let $\mathscr{S}$ be any sentence and let $\mathscr{S}_1, \ldots, \mathscr{S}_n$ be the propositional variables that appear in $\mathscr{S}$. Let v be any valuation. Then if $v(\mathscr{S}_i) = \mathrm{t}$ let $\mathscr{S}_i'$ be $\mathscr{S}_i$ while if $v(\mathscr{S}_i) = \mathrm{f}$ let $\mathscr{S}_i'$ be $\neg \mathscr{S}_i$. Also let $\mathscr{S}'$ be $\mathscr{S}$ if $v(\mathscr{S}) = \mathrm{t}$ and $\mathscr{S}'$ be $\neg \mathscr{S}$ if $v(\mathscr{S}) = \mathrm{f}$. Then we have that $\{\mathscr{S}_1', \ldots, \mathscr{S}_n'\} \vdash \mathscr{S}'$.

Proof This is by induction on the structure of $\mathscr{S}$. Let us assume that $\mathscr{S}$ has m occurrences of connectives.

In the case that $m = 0$ $\mathscr{S}$ is atomic and so consists of a single propositional variable. Clearly, if $v(\mathscr{S}) = \mathrm{t}$ then we have $\mathscr{S} \vdash \mathscr{S}$ and if $v(\mathscr{S}) = \mathrm{f}$ then $\mathscr{S}' = \neg\mathscr{S}$ and so $\neg\mathscr{S} \vdash \neg\mathscr{S}$. So the theorem holds for this simplest case. Now assume that it holds for any $j < m$. There are several cases and subcases to deal with.

(a) If $\mathscr{S}$ is of the form $\mathscr{T}_1 \to \mathscr{T}_2$ then $\mathscr{T}_1$ and $\mathscr{T}_2$ have less than m connectives and so by the assumption we have $\{\mathscr{S}_1', \ldots, \mathscr{S}_k'\} \vdash \mathscr{T}_1'$ and $\{\mathscr{S}_1', \ldots, \mathscr{S}_k'\} \vdash \mathscr{T}_2'$, where $\mathscr{S}_1', \ldots, \mathscr{S}_k'$ are the propositional variables in $\mathscr{S}$.

- (i) If $v(\mathscr{T}_1) = \mathrm{t}$ then $\mathscr{T}_1'$ is $\mathscr{T}_1$ and if $v(\mathscr{T}_2) = \mathrm{t}$ then $\mathscr{T}_2'$ is $\mathscr{T}_2$. Also we have $v(\mathscr{T}_1 \to \mathscr{T}_2) = \mathrm{t}$ and so $\mathscr{S}'$ is $\mathscr{T}_1 \to \mathscr{T}_2$. By the assumption, therefore, $\{\mathscr{S}_1', \ldots, \mathscr{S}_k'\} \vdash \mathscr{T}_2$ and since $\vdash \mathscr{T}_2 \to (\mathscr{T}_1 \to \mathscr{T}_2)$ we have, by modus ponens, that $\{\mathscr{S}_1', \ldots, \mathscr{S}_k'\} \vdash \mathscr{T}_1 \to \mathscr{T}_2$, that is $\{\mathscr{S}_1', \ldots, \mathscr{S}_k'\} \vdash \mathscr{S}'$.
- (ii) If $v(\mathscr{T}_1) = \mathrm{t}$ then $\mathscr{T}_1'$ is $\mathscr{T}_1$ and if $v(\mathscr{T}_2) = \mathrm{f}$ then $\mathscr{T}_2'$ is $\neg\mathscr{T}_2$. Also we have $v(\mathscr{T}_1 \to \mathscr{T}_2) = \mathrm{f}$ and so $\mathscr{S}'$ is $\neg(\mathscr{T}_1 \to \mathscr{T}_2)$. By the assumption, therefore, $\{\mathscr{S}_1', \ldots, \mathscr{S}_k'\} \vdash \neg\mathscr{T}_2$ and $\{\mathscr{S}_1', \ldots, \mathscr{S}_k'\} \vdash \mathscr{T}_1$ and since we can show that $\vdash \mathscr{T}_1 \to (\neg\mathscr{T}_2 \to \neg(\mathscr{T}_1 \to \mathscr{T}_2))$ we have, by modus ponens twice, that $\{\mathscr{S}_1', \ldots, \mathscr{S}_k'\} \vdash \neg(\mathscr{T}_1 \to \mathscr{T}_2)$, that is $\{\mathscr{S}_1', \ldots, \mathscr{S}_k'\} \vdash \mathscr{S}'$.
- (iii) If $v(\mathscr{T}_1) = \mathrm{f}$ then $\mathscr{T}_1'$ is $\neg\mathscr{T}_1$ and, whatever value v gives $\mathscr{T}_2$, we have $v(\mathscr{T}_1 \to \mathscr{T}_2) = \mathrm{t}$ and so $\mathscr{S}'$ is $\mathscr{T}_1 \to \mathscr{T}_2$. By the assumption, therefore, $\{\mathscr{S}_1', \ldots, \mathscr{S}_k'\} \vdash \neg\mathscr{T}_1$ and since $\vdash \neg\mathscr{T}_1 \to (\mathscr{T}_1 \to \mathscr{T}_2)$ we have, by modus ponens, that $\{\mathscr{S}_1', \ldots, \mathscr{S}_k'\} \vdash \mathscr{T}_1 \to \mathscr{T}_2$, that is $\{\mathscr{S}_1', \ldots, \mathscr{S}_k'\} \vdash \mathscr{S}'$.

(b) If $\mathscr{S}$ is of the form $\neg\mathscr{T}_1$ then $\mathscr{T}_1$ has less than n connectives and so $\{\mathscr{S}_1', \ldots, \mathscr{S}_k'\} \vdash \mathscr{T}_1'$, by the assumption.

- (i) If $v(\mathscr{T}_1) = \mathrm{t}$ then $\mathscr{T}_1'$ is $\mathscr{T}_1$ and $v(\mathscr{S}) = \mathrm{f}$ so $\mathscr{S}'$ is $\neg\mathscr{S}$. Since $\vdash \mathscr{T}_1 \to \neg\neg\mathscr{T}_1$ we have by modus ponens that $\{\mathscr{S}_1', \ldots, \mathscr{S}_k'\} \vdash \neg\neg\mathscr{T}_1$, that is $\{\mathscr{S}_1', \ldots, \mathscr{S}_k'\} \vdash \neg\mathscr{S}$, that is $\{\mathscr{S}_1', \ldots, \mathscr{S}_k'\} \vdash \mathscr{S}'$.
- (ii) If $v(\mathscr{T}_1) = \mathrm{f}$ then $\mathscr{T}_1'$ is $\neg\mathscr{T}_1$ and $v(\mathscr{S}) = \mathrm{t}$ so $\mathscr{S}'$ is $\mathscr{S}$. Therefore $\{\mathscr{S}_1', \ldots, \mathscr{S}_k'\} \vdash \neg\mathscr{T}_1$, that is $\{\mathscr{S}_1', \ldots, \mathscr{S}_k'\} \vdash \mathscr{S}'$.

With that we have covered all cases and, by induction on n, the proof is complete. ■

Theorem

If $\vDash \mathscr{S}$ then $\vdash \mathscr{S}$, for any sentence $\mathscr{S}$.

Proof Assume that $\vDash \mathscr{S}$. Let $\mathscr{S}_1, \ldots, \mathscr{S}_n$ be the propositional variables in $\mathscr{S}$.

By Lemma 2.2 we know that, for any valuation, $\{\mathscr{S}_1', \ldots, \mathscr{S}_n'\} \vdash \mathscr{S}$. Note here that $\mathscr{S}'$ is $\mathscr{S}$ since $\models \mathscr{S}$, so for any valuation $\mathscr{S}$ is true. Therefore we have that $\{\mathscr{S}_1', \ldots, \mathscr{S}_{n-1}', \mathscr{S}_n\} \vdash \mathscr{S}$ and $\{\mathscr{S}_1', \ldots, \mathscr{S}_{n-1}', \neg \mathscr{S}_n\} \vdash \mathscr{S}$.

Using the deduction theorem we have $\{\mathscr{S}_1', \ldots, \mathscr{S}_{n-1}'\} \vdash \mathscr{S}_n \rightarrow \mathscr{S}$ and $\{\mathscr{S}_1', \ldots, \mathscr{S}_{n-1}'\} \vdash \neg \mathscr{S}_n \rightarrow \mathscr{S}$. Since we can show $\vdash (\mathscr{S}_n \rightarrow \mathscr{S}) \rightarrow ((\neg \mathscr{S}_n \rightarrow \mathscr{S}) \rightarrow \mathscr{S})$ we have, using modus ponens twice, $\{\mathscr{S}_1', \ldots, \mathscr{S}_{n-1}'\} \vdash \mathscr{S}$. By repeating this process $n - 1$ more times we have $\vdash \mathscr{S}$ as required. ■

EXAMPLE 2.3

As an example of how this proof works, consider the case in which $\mathscr{S}$ is $(p \rightarrow (q \rightarrow p))$. Then, by Lemma 2.2 we have $\{p', q'\} \vdash (p \rightarrow (q \rightarrow p))$. Writing this out in full we have

$$\{p, q\} \vdash (p \rightarrow (q \rightarrow p))$$

and

$$\{p, \neg q\} \vdash (p \rightarrow (q \rightarrow p))$$

and

$$\{\neg p, q\} \vdash (p \rightarrow (q \rightarrow p))$$

and

$$\{\neg p, \neg q\} \vdash (p \rightarrow (q \rightarrow p)).$$

So by the deduction theorem we have

$$\{p\} \vdash (q \rightarrow (p \rightarrow (q \rightarrow p))) \text{ and } \{p\} \vdash (\neg q \rightarrow (p \rightarrow (q \rightarrow p)))$$

and

$$\{\neg p\} \vdash (q \rightarrow (p \rightarrow (q \rightarrow p))) \text{ and } \{\neg p\} \vdash (\neg q \rightarrow (p \rightarrow (q \rightarrow p)))$$

The first two of these can be combined twice with

$$\vdash (q \rightarrow (p \rightarrow (q \rightarrow p))) \rightarrow ((\neg q \rightarrow (p \rightarrow (q \rightarrow p))) \rightarrow (p \rightarrow (q \rightarrow p)))$$

using modus ponens to give

$$\{p\} \vdash (p \rightarrow (q \rightarrow p))$$

and similarly for the second two to give

$$\{\neg p\} \vdash (p \rightarrow (q \rightarrow p))$$

This can then be repeated to give

$$\vdash (p \rightarrow (q \rightarrow p))$$

SUMMARY

- A logical *language* is defined by giving its *alphabet* and its *grammar* – a set of rules that says which strings of symbols from the alphabet are well-formed *sentences*. We gave the alphabet and grammar of *propositional logic*. The property of being a sentence is *decidable* and there is a computationally effective method of *enumerating* the sentences of the language. We demonstrated this by giving SML and Prolog programs for decidability and enumeration.
- An important part of logic is the study of *arguments* – of which conclusions can correctly be said to follow from given premises. The validity of an argument is judged from its *form* not its meaning. A valid argument can have a false conclusion – provided its premises are false – but a valid argument applied to true premises always gives a true conclusion. In classical propositional logic the validity of argument forms can be decided by *truth-tables*.
- Meaning is given to a language via a set of rules that enable you to calculate what the sentences of the language *denote*. Languages in which you can calculate the truth-value of a sentence from the truth-values of its subsentences are called *truth-functional*. Classical propositional logic is truth-functional and the calculation rules are given by truth-tables for each of the connectives. A truth-table can be constructed for any given sentence by listing all the ways in which truth-values can be assigned to its constituent propositional variables and, using the truth-tables for the connectives, calculating the value of the sentence in each case. An assignment of truth-values to propositional variables is called a *valuation*. A formula that is true in every valuation is said to be *valid*. Valid formulas in propositional logic are called *tautologies*.
- A set of sentences G *entails* a sentence $\mathscr{S}$, written $G \models \mathscr{S}$, if every valuation that makes all the sentences in G true makes $\mathscr{S}$ true also. If no valuation makes all the sentences in G true then G is said to be *inconsistent*.
- A *formal system* enables the validity of arguments in a

language to be decided without reference to the notions of true and false. As well as the alphabet and grammar of the language, a formal system has *axioms* and *rules of inference.* A *proof* in such a system is a sequence of sentences, each of which is either an axiom or is derived from earlier members of the sequence by using the rules of inference. The final sentence of such a sequence is said to be a *theorem* of the system. It can be shown that the formal system for classical propositional logic given in this chapter is *sound* and *complete* in the sense that all its theorems are tautologies and all tautologies are among its theorems.

3 Extending the Language

3.1 Informal predicate calculus

The simplicity of the propositional calculus is also its worst shortcoming. A simple argument such as

All men are mortal

Socrates is a man

therefore

Socrates is mortal

is of the form

A

M

therefore

D

or, in propositional logic

$$(A \wedge M) \rightarrow D$$

but this, as you can verify, is not a valid argument; the truth-table does not give 't' for every valuation. However, we would obviously like to characterize such arguments as valid. To do this we have to look at the internal structure of the propositions.

'All men are mortal' has the natural language form 'if x is a man then x is mortal, for all x'. We are going to define a language called predicate calculus in which this can be written formally as $\forall x(M(x) \rightarrow D(x))$.

Here, x is a **variable** which is understood to stand for any object, $\forall$ is a **universal quantifier** ('for all ...') and M and D are one-place **predicates**. That is, they express properties that the objects which are their arguments have.

If we write 's' for Socrates we have $M(s)$ and $D(s)$ for the other two sentences in the argument. Now we can show the argument to be valid.

$\forall x(M(x) \rightarrow D(x))$ says that, for any x, $(M(x) \rightarrow D(x))$ so it is certainly true to say $(M(s) \rightarrow D(s))$. But we are given $M(s)$, so by propositional logic we can deduce $D(s)$.

We now give the formal definition of the language.

3.1.1 Grammatical definitions

Definition 3.1: **A language of predicate calculus based on $\langle \mathsf{P}, \mathsf{N}, \mathsf{F} \rangle$**, written $L(\langle \mathsf{P}, \mathsf{N}, \mathsf{F} \rangle)$, is defined by the following.

(a) The **alphabet** which has the seven components:

(i) a set of connective symbols $\{\neg, \rightarrow, \wedge, \vee, \leftrightarrow\}$,
(ii) a set of quantifier symbols $\{\forall, \exists\}$,
(iii) a set of punctuation symbols $\{(,), ,\}$,
(iv) a set P of sets P_n of n-ary predicate symbols for each $n \geqslant 0$,
(v) a set N of names $\{a, b, c, \ldots\}$,
(vi) a set F of sets F_n of n-ary function symbols $\{f, g, h, \ldots\}$ for each $n \geqslant 0$.

We call P, N and F the **parameters** of the language.

Each language also contains (vii) a set V of variables $\{x_1, x_2, \ldots\}$. However, since this set is included in all languages there is no need to include it in the parameters that define the language.

(b) The **grammar**, which defines $L(\langle \mathsf{P}, \mathsf{N}, \mathsf{F} \rangle)$, the set of **sentences based on $\langle \mathsf{P}, \mathsf{N}, \mathsf{F} \rangle$**. It is given here informally by describing several kinds of objects which go to make up sentences (and as before we shall not mention the basis of the sentence and its parts explicitly unless it is not clear from the context what we are referring to).

A **term based on $\langle \mathsf{P}, \mathsf{N}, \mathsf{F} \rangle$** is a variable, or a name, or a function symbol followed in parentheses, and separated by commas, by the correct number of terms for the function symbol's arity.

For example x, a and $f(x, a, g(y, z))$ are all terms if f has arity 3, g has arity 2, a is a name and x, y and z are variables.

A **formula based on** $\langle \mathsf{P}, \mathsf{N}, \mathsf{F} \rangle$ is of the form

$$\neg\mathscr{F}, \text{ or } (\mathscr{F} * \mathscr{G}), \text{ or } Qv\mathscr{F}, \text{ or } P(t_1, t_2, \ldots, t_n),$$

where $*$ is one of the binary logical connectives, $\mathscr{F}$ and $\mathscr{G}$ are formulas, called **subformulas** of the formula, Q is a quantifier, v is a variable and $P(t_1, t_2, \ldots, t_n)$ is an **atomic formula** consisting of a predicate symbol followed in parentheses, and separated by commas, by the correct number of terms for its arity.

In the case $Qv\mathscr{F}$ where $\mathscr{F}$ contains no quantifiers we say that any occurrence of v in $\mathscr{F}$ is **bound by Q in $\mathscr{F}$**, and $\mathscr{F}$ is the **scope** of Q. If an occurrence of a variable v is bound by a quantifier then it is a **bound occurrence** and we say that the quantifier **binds** v. In the case of $Qv\mathscr{F}$ where $\mathscr{F}$ does contain quantifiers then any occurrences of v in $\mathscr{F}$ which are not bound by any other quantifier in $\mathscr{F}$ are bound by Q. In this case the scope of Q consists of all subformulas of $\mathscr{F}$ that are not in the scope of a quantifier other than Q that binds v.

Note that no occurrence of a variable can be bound by more than one quantifier. If an occurrence of a variable v is not bound by any quantifier then that occurrence is a **free occurrence**.

A **sentence based on** $\langle \mathsf{P}, \mathsf{N}, \mathsf{F} \rangle$ is a formula in which no free variables occur. An atomic formula with no free variables is an **atomic sentence**.

For example, if a is a name and x is a variable then $f(x, a)$ is a term, provided that f is a function symbol of arity 2, $(A \rightarrow B(x, f(y, a)))$ is a formula in which x and y are free, $\forall x \exists y(A \rightarrow B(x, f(y, a)))$ is a sentence and $R(a, b, c)$ is an atomic sentence if A, B and R are predicate symbols.

One final definition in this section is of the notion of a term t being **free for** a variable v. This essentially means that no variables of t are 'captured' when t is substituted for v in a formula $\mathscr{F}$.

Definition 3.2: If $\mathscr{F}$ is a formula and v is a free variable of $\mathscr{F}$ then the term t is **free for** v in $\mathscr{F}$ iff there is no variable u in t such that v appears within the scope of a quantifier that binds u in $\mathscr{F}$.

For example, if $\forall x_1 A(x_1, x_2)$ is the formula in question, then $f(x_1, x_3)$ is not free for x_2. Since x_2 appears within the scope, namely $A(x_1, x_2)$, of a quantifier that binds x_1, the x_1 in $f(x_1, x_3)$ is captured by the quantifier. However, $f(x_2, x_3)$ is free for x_2 in the same formula since the only quantifier binds x_1 and this variable does not appear in $f(x_2, x_3)$.

Now we can extend the SML declarations that define the language in the previous section. The extensions that we need are to do with the terms, and the addition of predicates and quantifiers to the language. First, we have to decide on how to represent the terms. From the definition a term is a variable or a name or a function whose parameter places are filled with the correct number of terms for its arity. So the following SML datatype can be used:

```
datatype TERM = Var of string
              | Name of string
              | app of (string * TERM list)
              | empty;
```

So, the term which is the name 'a' will be represented as `Name "a"`; the variable 'x' will be represented as `Var "x"`; the application of a function to its arguments as in 'g(a, b)' will be represented by `app("g", [Name "a",Name "b"])` and so on.

Having represented the terms we can extend the declaration of SENT, for (roughly, since we have not yet parsed them) the sentences:

```
datatype SENT = Prop of string
              | Not of SENT
              | And of (SENT * SENT)
              | Or of (SENT * SENT)
              | Imp of (SENT * SENT)
              | Eq of (SENT * SENT)
              | Forall of (TERM * SENT)
              | Exists of (TERM * SENT)
              | Pred of (string * TERM list);
```

3.1.2 Interpretations

We now go on to define the meaning of the sentences and formulas of the language. This is done by introducing the notion of interpretation:

Definition 3.3: An **interpretation** of a language $L(\langle \mathsf{P}, \mathsf{N}, \mathsf{F} \rangle)$ – and we shall shorten this to just L in this section – over a universe U is a pair of functions

$$I_U^L = \langle \Phi_U, \Psi_U \rangle^L$$

where

- U is a non-empty set of objects, which the terms refer to, called the **universe**;
- Ψ_U: $P_n \to \mathcal{P} U^n$ maps, for each P_n in P, the n-ary predicate symbols to n-ary relations over U;
- Φ_U: $\mathsf{N} \to U$ maps the names to objects in the universe;
- Φ_U: $F_n \to (U^n \to U)$ maps, for each F_n in F, the n-ary function symbols to n-ary functions over the universe.

Ψ_U maps n-ary predicate symbols to the relations which they denote. We think of a relation as given by the set of tuples of objects from U which are in the relation to one another (which you may see called its **extension**). For instance, in mathematics the symbol '$<$' as a relation over the universe of the positive integers $\{1, 2, 3, \ldots\}$ has, as its extension, the set of ordered pairs $\{(1, 2), (1, 3), (2, 3), \ldots\}$ because 1 is less than 2, 1 is less than 3, 2 is less than 3 and so on.

Note that Φ_U is **overloaded**, that is the symbol stands for two operations. However, this is not ambiguous because, when it is applied, we can always tell from its argument which operation to use.

To make such a Φ_U complete, each variable must also denote an object in U. This is done by considering Φ_U to be, in fact, a set of functions, the elements of which we call **assignments** and denote by ϕ_U, All the assignments in the set Φ_U agree on the values of the names and function symbols. Hence, on the names and function symbols, we are justified in thinking of just one assignment. This is why we allowed the abuse of notation above that made Φ_U look as if it were a single function. The assignments differ on what they map the variables to and there are enough of them to allow a given variable to be mapped to each of the objects in the universe U.

Finally, we have to extend each ϕ_U so that, when applied to an element from the set of terms T, it produces as its value an object from U. So, we define

$$\phi_U(g(t_1, \ldots, t_m)) = \phi_U(g)(\phi_U(t_1), \ldots, \phi_U(t_m))$$

for each $g \in F_m$, $t_i \in T$.

The name 'assignment' varies between authors. Some (Hamilton (1978) for example) use the word 'valuation' for Φ (and in the way we used it previously, which is rather confusing) and Mendelson (1987) uses 'sequence' for part of it. Some use no name at all. Hodges (1977) seems to use 'assignment' too. Our presentation is governed by the wish to keep the mechanism for giving a meaning to predicate symbols distinct from that for giving a meaning to terms. This means that the predicate case collapses immediately to the propositional case, where we called Ψ a valuation, in a language with no terms.

We are now in a position to give a hierarchy by which the truth of a sentence in a particular language L can be determined.

Definition 3.4: An assignment ϕ_U satisfies a formula in an interpretation I_U^L according to the following where $\mathscr{F}$ and $\mathscr{G}$ are any formulas:

(1) for any predicate $Q \in P_n$, ϕ_U satisfies $Q(t_1, \ldots, t_n)$ for any $t_i \in T$, iff $(\phi_U(t_1), \ldots, \phi_U(t_n)) \in \Psi_U(Q)$.

(2) ϕ_U satisfies $\neg\mathscr{F}$ iff ϕ_U does not satisfy $\mathscr{F}$.

(3) ϕ_U satisfies $(\mathscr{F} \wedge \mathscr{G})$ iff ϕ_U satisfies both $\mathscr{F}$ and $\mathscr{G}$.

(4) ϕ_U satisfies $(\mathscr{F} \vee \mathscr{G})$ iff ϕ_U satisfies either $\mathscr{F}$ or $\mathscr{G}$, or both.

(5) ϕ_U satisfies $(\mathscr{F} \rightarrow \mathscr{G})$ iff ϕ_U satisfies $(\neg\mathscr{F} \vee \mathscr{G})$.

(6) ϕ_U satisfies $(\mathscr{F} \leftrightarrow \mathscr{G})$ iff ϕ_U satisfies $(\mathscr{F} \wedge \mathscr{G}) \vee (\neg\mathscr{F} \wedge \neg\mathscr{G})$.

(7) ϕ_U satisfies $\exists x\mathscr{F}$ iff there is an element a of U such that there is some assignment ϕ_U' from I_U^L, which is the same as ϕ_U except perhaps at x where $\phi_U'(x) = a$, which satisfies $\mathscr{F}$.

(8) ϕ_U satisfies $\forall x\mathscr{F}$ iff ϕ_U satisfies $\neg\exists x\neg\mathscr{F}$. This is the same as saying that there is no element a of U such that there is some assignment ϕ_U' from I_U^L which is the same as ϕ_U, except perhaps at x where $\phi_U'(x) = a$, which does not satisfy $\mathscr{F}$. This, in turn, says that for every element a in U all assignments ϕ_U', which are the same as ϕ_U, except perhaps at x where $\phi_U'(x) = a$, satisfy $\mathscr{F}$.

The fact that variables are only allowed to range over members of the universe leads to other names for the language that we are studying. In full we should call it a **first-order predicate** language. In the same vein, the propositional language, where quantifiers ranged over nothing (since there were none) might be called a zero-order predicate language. So, the order of the language refers to the 'level' of the objects up to and including which the quantifiers can range over. You can think of the level as being the place that the objects come in the series which starts

nothing	zeroth-order
the universe	first-order
relations over the universe (that is predicates)	second-order
relations over relations	third-order
⋮	⋮

We shall use the term 'predicate logic' or 'predicate calculus' to refer only to the first-order case, so there is no ambiguity. Sometimes we may talk of 'first-order logic' or 'first-order language' when we refer to this same level.

Definition 3.5: A formula is **true** in an interpretation I_U^L iff all the assignments in I_U^L satisfy it. A formula is **valid** iff it is true in every interpretation of its language L. Conversely, a formula is **false** in an interpretation I_U^L iff no assignment in I_U^L satisfies it and it is **unsatisfiable** iff it is false in every interpretation. Formulas that are not unsatisfiable are **satisfiable**.

Note that in the propositional case, since there are no terms, the assignment functions are redundant and then an interpretation collapses to what we called a valuation (as long as we equate the propositions with nullary predicates).

Note that the definition implies that if we wish to see whether a

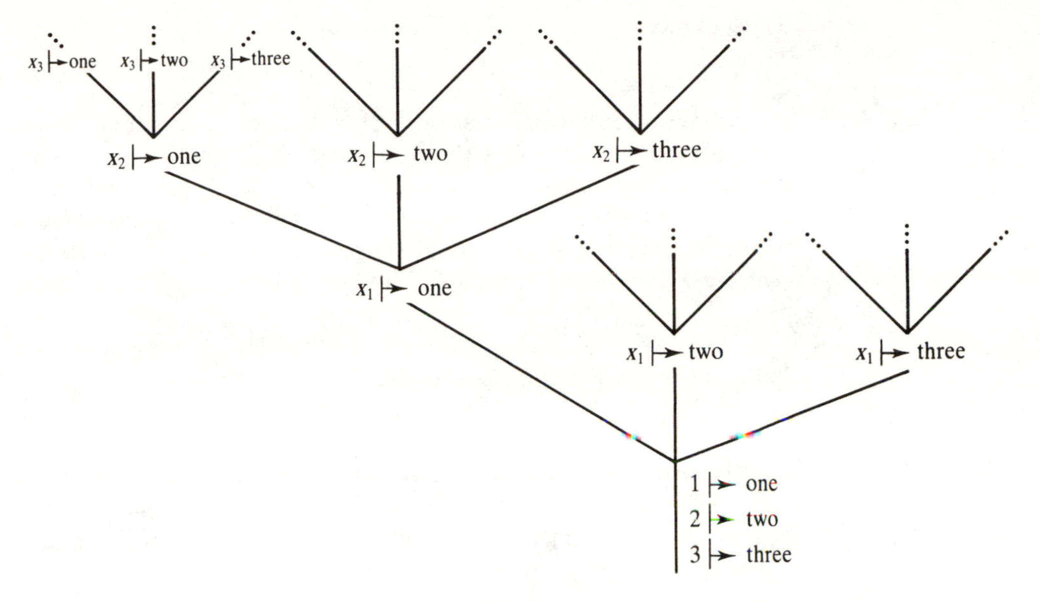

Figure 3.1.

formula is true in some interpretation then we have to show that it is satisfied by all assignments in the interpretation. This, clearly, may involve a lot of work (especially if the number of assignments is infinite). However, it can be proved that a formula is satisfied by all assignments in an interpretation iff it is satisfied by at least one assignment in the interpretation, which means that the amount of work required to see whether a formula is true in an interpretation can be vastly reduced.

3.1.3 Examples on interpretations

(1) The sentence $\exists x_1(1 < x_1)$ is true in the interpretation which makes '1' denote the number one and '$<$', which is an infixed predicate symbol in the formal language, denote the less-than relation over the integers {one, two, three}. To see this in a graphical way imagine all the possible assignments in this interpretation displayed as a tree (Figure 3.1). The common part of all the assignments maps 1 to one, 2 to two, 3 to three. This is shown on the trunk of the tree. The branches then show all the possible extensions which give all the possible assignments.

We can see that for an arbitrary assignment (path up the tree) there is an assignment that is the same everywhere except possibly at x_1, where it maps x_1 to two. This means that the assignment satisfies the existential sentence and so it meets the condition for being satisfied by any assignment and so it is true under the interpretation.

(2) The sentence

$$1 < 2 \rightarrow \exists x(x < 2)$$

is valid. We can think of this as saying that 'in those assignments where the left-hand side is satisfied, the right-hand side is too'. This is the case since any assignment which satisfies $1 < 2$ will clearly (trivially) satisfy $\exists x(x < 2)$. This argument then extends to any interpretation, which might be clearer if we use a less suggestive language and write the sentence as

$$P(a, b) \rightarrow \exists x P(x, b)$$

(3) In the same interpretation as above

$$\exists x(x < 1)$$

is a false sentence. This is so because, for an arbitrary assignment, there is no assignment which is the same, except perhaps at x_1, such that x_1 is mapped to an element of the universe which is in the relation less-than to 1.

(4) An example of an unsatisfiable sentence is $1 < 2 \wedge \forall x \neg(x < 2)$.

(5) The sentence

$$\forall x \forall y \forall z(x < y \wedge y < z \rightarrow x < z)$$

is satisfiable. It is true if '$<$' is less than over the natural numbers, and it is false if '$<$' denotes 'is the square root of' over the same universe, since, for instance, we have $2 < 4$ and $4 < 16$ but not $2 < 16$.

In general, to prove validity, we must show that an arbitrary assignment in an arbitrary interpretation satisfies the sentence concerned.

On the other hand, to prove a sentence non-valid we must be ingenious enough actually to construct an interpretation in which there is an assignment which does not satisfy the sentence.

(6) For a more extended example we consider a language with parameters as follows.

The set of predicate symbols is $\{M, S\}$, both symbols having arity 2; the set of function symbols is $\{f', m'\}$, both with arity 1; the set of names is $\{a, b, c, d, e, f, g\}$. This defines the language so that the following are all sentences:

$$M(a, b)$$
$$M(a, b) \rightarrow M(b, a)$$
$$\neg M(c, d)$$
$$\forall x M(a, x)$$
$$\exists x M(a, x)$$
$$M(f'(a), m'(c))$$

However, even though we know that these are sentences, we still have no way of knowing their meaning. For this we need an interpretation:

$I = \langle \Phi_U, \Psi_U \rangle$ where $U = \{$Albert, Brian, Cynthia, Dougal, Ermintrude, Florence, Georgina$\}$

$\Psi_U(M) = \{$(Cynthia, Brian), (Brain, Cynthia)$\}$

$\Psi_U(S) = \{$(Ermintrude, Dougal), (Dougal, Ermintrude)$\}$

$\Phi_U(f') = \{$(Cynthia$\mapsto$Albert, Florence$\mapsto$Brian)$\}$

$\Phi_U(m') = \{$Florence$\mapsto$Cynthia, Ermintrude$\mapsto$Florence, Dougal$\mapsto$Florence$\}$

$\Phi_U(a) =$ Albert, $\Phi_U(b) =$ Brian, $\Phi_U(c) =$ Cynthia, $\Phi_U(d) =$ Dougal, $\Phi_U(e) =$ Ermintrude, $\Phi_U(f) =$ Florence, $\Phi_U(g) =$ Georgina

Now we can work out the truth-values of some of the sentences above. For example, for $M(a, b)$, $(\Phi_U(a), \Phi_U(b)) =$ (Albert,Brian) $\notin$ $\{$(Cynthia,Brian), (Brian,Cynthia)$\}$ $= \Psi_U(M)$, so $M(a, b)$ is not true.

(7) As another example, we can give a different interpretation, J, to the same language as above. We define J as follows:

$J = \langle \Phi_V, \Psi_V \rangle$ where $V = \{$Anne, Bill, Charles, Diana, Elizabeth, Fred, Gertrude$\}$

$\Psi_V(M) = \{$(Anne, Bill), (Bill, Anne), (Fred, Gertrude), (Gertrude, Fred)$\}$

$\Psi_V(S) = \{$(Charles, Diana), (Diana, Charles), (Bill, Elizabeth), (Elizabeth, Bill)$\}$

$\Phi_V(f') = \{$Diana$\mapsto$Bill, Charles$\mapsto$Bill, Bill$\mapsto$Fred, Elizabeth$\mapsto$Fred$\}$

$\Phi_V(m') = \{$Elizabeth$\mapsto$Gertrude, Bill$\mapsto$Gertrude, Charles$\mapsto$Anne, Diana$\mapsto$Anne$\}$

$\Phi_V(a) =$ Anne, $\Phi_V(b) =$ Bill, $\Phi_V(c) =$ Charles, $\Phi_V(d) =$ Diana ...

We have, for $M(a, b)$, that $(\Phi_V(a), \Phi_V(b)) =$ (Anne,Bill) $\in$ $\{$(Anne, Bill), (Bill, Anne), (Fred, Gertrude), (Gertrude, Fred)$\}$ $= \Psi_V(M)$, so $M(a, b)$ is true.

Exercise 3.1

(a) Work out whether the following are true in each of the interpretations I and J given in the previous section:

(i) $M(a, c)$

(ii) $M(a, b) \rightarrow M(b, a)$

(iii) $S(e, d)$

(iv) $S(e, d) \rightarrow S(d, e)$
(v) $\forall x \forall y(M(x, y) \rightarrow M(y, x))$
(vi) $\exists z M(f'(c), b)$

(b) From your knowledge of family relationships and the legalities of marriage choose suitable denotations for the function symbols f' and m', and the predicate symbols M and S in the examples above. Draw the family trees that the interpretations I and J describe.

(c) A first-order language L has the usual connectives and punctuation symbols, together with variable symbols x and y, a constant symbol a, a unary function symbol t and two binary predicate symbols $=$ and $<$, both used in infixed form.

The domain of an interpretation I for L is the set of integers $D = \{0, 1, 2\}$. The denotation of a is the number 0. The denotation of t is the function from D to D that, for any $n \in D$, maps n to the remainder on dividing $(n + 2)$ by 3. The denotation of $<$ is the binary relation on D whose extension is $\{\langle 0, 1\rangle, \langle 1, 2\rangle, \langle 2, 0\rangle\}$ and the denotation of $=$ is the usual identity relation over integers.

For each of the following strings of symbols say whether or not it is a formula of L. If it is, say whether the formula is true, satisfiable, or false in I. Justify your statement briefly. The first one has been completed for you as an example.

(i) $t(t(a)) = a$ is a formula of L and is false, because $t(a)$ denotes the number 2, so $t(t(a))$ denotes the number $(2 + 2) \bmod 3 = 1$ and 1 is not the same number as 0.
(ii) $t(a) < a$ is ....
(iii) $\exists x(x > t(x))$ is ....
(iv) $\forall x(t(t(t(x))) = x)$ is ....
(v) $\forall x \exists y(t(x) < t(y))$ is ....
(vi) $\forall x(t(x) < t(y))$ is ....
(vii) $\forall x((t(x) < t(y)) \rightarrow \neg(x = y))$ is ....

3.2 FDS for predicate calculus

We have already seen an alphabet and (informal) grammar for this system. The following is a possible collection of **axiom schemas**.

Let $\mathscr{F}$, $\mathscr{G}$ and $\mathscr{H}$ be formulas of the language. Then:

(P1) $(\mathscr{F} \rightarrow (\mathscr{G} \rightarrow \mathscr{F}))$

(P2) $((\mathscr{F} \rightarrow (\mathscr{G} \rightarrow \mathscr{H})) \rightarrow ((\mathscr{F} \rightarrow \mathscr{G}) \rightarrow (\mathscr{F} \rightarrow \mathscr{H})))$

(P3) $(((\neg \mathscr{F}) \rightarrow (\neg \mathscr{G})) \rightarrow (\mathscr{G} \rightarrow \mathscr{F}))$

(P4) $(\forall x_i \mathscr{F} \rightarrow \mathscr{F})$, if x_i does not occur free in $\mathscr{F}$

(P5) $(\forall x_i \mathscr{F} \rightarrow \mathscr{F}[t/x_i])$, if $\mathscr{F}$ is a formula of the language in which x_i may appear free and t is free for x_i in $\mathscr{F}$

(P6) $(\forall x_i(\mathscr{F} \rightarrow \mathscr{G}) \rightarrow (\mathscr{F} \rightarrow \forall x_i \mathscr{G}))$, if x_i does not occur free in $\mathscr{F}$

Rules of deduction

(1) Modus ponens (MP), from $\mathscr{F}$ and $(\mathscr{F} \rightarrow \mathscr{G})$ deduce $\mathscr{G}$ where $\mathscr{F}$ and $\mathscr{G}$ are any formulas of the language.

(2) Generalization, from $\mathscr{F}$ deduce $\forall x_i \mathscr{F}$, where $\mathscr{F}$ is any formula of the language.

Note that since the rules and axioms given above include those of the propositional calculus it is clear that any sentence that is a theorem of propositional calculus is also a theorem of predicate calculus.

One important fact about this new logic is that the deduction theorem does not go through quite as before. This time we need an extra condition in the premise of the theorem.

Theorem

Let $\mathscr{S}$ and $\mathscr{T}$ be any formulas and let G be any set of formulas.

Then, if $G, \mathscr{S} \vdash \mathscr{T}$, and no use was made of generalization involving a free variable of $\mathscr{S}$, it is the case that $G \vdash \mathscr{S} \rightarrow \mathscr{T}$.

It can be shown that predicate calculus has the same soundness and completeness properties that we proved for propositional calculus, although in this case we shall not give the proofs since they are even more involved than was the case for propositional calculus, though they follow the same lines.

One property that predicate calculus does not have is that of decidability. That is, the question 'Is this sentence valid or not?' cannot be answered if we use any method which, essentially, can be programmed. We say that predicate calculus is **undecidable** (though we will be able to refine this statement later on). The lack of an answer does not depend on how clever the method is, or how long we set it to work for. It is simply that for some sentences there is no answer.

3.2.1 Deciding on proofhood

Even though there are bounds to what is possible computationally, we can extend some of the programs that we had in the sections on propositional logic to the predicate case since, although deciding on validity is not possible in general, deciding on proofhood is. This is because a proof is simply a finite sequence of finite sentences, each of which has to have some decidable property that we can check.

We saw above how the language can be extended to include terms, quantifiers and predicates. In order to be able to check a sequence of formulas to see whether it satisfies the definition of a proof in predicate logic, as we did for propositional logic, we first need some programs for checking for free variables, deciding on axiomhood and so on. The first function we define tests to see whether a given variable occurs in a given term and of course the program follows the definition of the structure of a term:

```
infix occurs_in;
fun v occurs_in ((Name n)::t) = false orelse (v occurs_in t)
  | v occurs_in ((Var w)::t) = (v=w) orelse (v occurs_in t)
  | v occurs_in (app(f,terms)::t) = (v occurs_in terms) orelse (v occurs_in t);

infix free_in;
fun v free_in (And(s,t)) = (v free_in s) orelse (v free_in t)
  | v free_in (Or(s,t)) = (v free_in s) orelse (v free_in t)
  | v free_in (Imp(s,t)) = (v free_in s) orelse (v free_in t)
  | v free_in (Eq(s,t)) = (v free_in s) orelse (v free_in t)
  | v free_in (Not(s,t)) = (v free_in s) orelse (v free_in t)
  | v free_in (Forall(Var y,s)) = if v=Var y
                                  then false
                                  else v free_in s
  | v free_in (Exists(Var y,s)) = if v=Var y
                                  then false
                                  else v free_in s
  | v free_in (Pred(p,terms)) = v occurs_in terms
  | v free_in (Prop p) = false;
```

This may look rather complicated but in fact the computation, as before, simply follows the structure of the sentence being tested. As we can see, for each of the connectives we just test for the variables being free in each subsentence. For the quantifiers, if the bound variable is the same as the variable then clearly it is not free. The only interesting case, really, is the atomic sentence of the form $P(t_1,\ldots,t_n)$. Here we just test to see whether the variable occurs in any of t_1–t_n. Finally, if the sentence is propositional then clearly the variable does not occur, so it does not occur free.

In order to test for axiomhood we also need a test to see whether a term is free for a variable in a sentence, but we assume that we have such a function, which we call free_for (and leave its development as an exercise).

We can now go on to give functions for the extended notion of axiom schemas:

```
fun axiom1 (Imp(S,Imp(T,S'))) = S=S'
  | axiom1 s = false;

fun axiom2 (Imp(Imp(S,Imp(T,R)),Imp(Imp(S',T'),Imp(S'',R')))) =
              (S=S') andalso (T=T') andalso (R=R') andalso (S=S'')
  | axiom2 s = false;

fun axiom3 (Imp(Imp(Not S,Not T),Imp(T',S'))) = (S=S') andalso (T=T')
  | axiom3 s = false;

fun axiom4 (Imp(Forall(x,S),S')) = not(x free_in S) andalso (S=S')
  | axiom4 _ = false;
```

```
fun axiom5 (Imp(Forall(x,S),S')) =
                        let val (ok,t) = forms_match (x,S) S' in
                          if ok then t free_for (x,S) else false
                        end
  | axiom5 _ = false;

fun axiom6 (Imp(Forall(x,Imp(S,T)),Imp(S',Forall(x',T')))) =
              S = S' andalso T = T'andalso x = x' andalso not(x free_in S)
  | axiom6 _ = false;

fun axiom s = axiom1 s orelse axiom2 s orelse axiom3 s
                 orelse axiom4 s orelse axiom5 s orelse axiom6 s;
```

The function forms_match is again left as an exercise, and it essentially takes two sentences and a variable and returns true if the sentences match up to occurrences of the variable and also the term that appears in place of the variable where the variable does not itself appear. For instance we have that forms_match applied to the formulas $P(x)$ and $P(f(a))$ and the variable x returns true, since the sentences are the same up to the occurrence of x, together with $f(a)$, since that is the term that appears in the place of x. If we apply it to $Q(x, x)$ and $Q(f(b), f(a))$ then it returns false since the formulas do not match because the terms in place of x are not the same.

The final part we need is to check for a correct use of the rule of generalization. This means that if a formula of the form $\forall x \mathcal{S}$ appears in the sequence of formulas that we are testing for proofhood then we also need to see whether a sentence of the form $\mathcal{S}$ appears before it in the sequence. The following SML function does just that:

```
fun generalization l (Forall(x,S)) = S memberof l
  | generalization _ _ = false;
```

We can then adapt and extend the function proof as given before to give a proof checker for predicate logic too. This is again left as an exercise.

Exercise 3.2

(a) Which of the following statements are correct and which incorrect?
In the formula $\forall x_1 A(x_2)$

(i) x_2 is free for x_2
(ii) x_2 is free for x_1
(iii) x_1 is free for x_2
(iv) x_7 is free for x_2
(v) $f(x_1, x_7)$ is free for x_2

(b) In the sentences and formulas below, is the term $f(x_1, x_3)$ free for x_2 or not?

(i) $\forall x_2(A(x_1, x_2) \rightarrow A(x_2, a))$
(ii) $\forall x_1 \forall x_3(A(x_1, x_2) \rightarrow B(x_3))$

(iii) $\forall x_2 A(f(x_2)) \rightarrow \forall x_3 B(x_1, x_2, x_3)$
(iv) $\neg\neg A(x_2) \wedge \forall x_5 A(x_2)$
(v) $\forall x_2 A(x_2)$

(c) Show that
(i) x_i is free for x_i in any formula;
(ii) any term is free for x_i in a formula $\mathscr{S}$ if $\mathscr{S}$ contains no free occurrences of x_i.

(d) Write an SML or Prolog program which tests whether a given term is free for a given variable in a given formula.

(e) Construct proofs of the following theorems of predicate calculus:
(i) $\forall x_1 \forall x_2 A \vdash \forall x_2 \forall x_1 A$
(ii) $\forall x_1 \forall x_2 A(x_1, x_2) \vdash \forall x_1 A(x_1, x_1)$

(f) The word 'theorem' is used in two different ways in Chapter 2. Give one example of each of these two uses of 'theorem', from Sections 2.5 or 2.6.

(g) Write an SML or Prolog program which implements the forms_match function needed for this section.

(h) Adapt and extend the program for proof in the previous chapter so that it works on proofs in predicate logic too.

3.3 Historical discussion

The undecidability of predicate logic is one of the fundamental results for computer science as well as for logic and it links our subject with the work of some of the greatest mathematicians and logicians of the modern era. First, the German mathematician David Hilbert, who was celebrated as the greatest mathematician of his time, being something of an all-rounder, proposed 23 problems at a congress of mathematicians from all over the world in 1900 (see Reid (1970) for his biography). He stated in the introduction to these problems that he was convinced that every mathematical problem could be answered, either with a solution or with a proof that it had no solution. He even referred to 'this axiom of solvability'. The problems that he listed (and, even though there were 23, he only described 10 in the lecture, since the talk was considered long enough already) led, as he had hoped when he propounded them, to some of the greatest discoveries in mathematics.

Unfortunately for Hilbert, as we now know and have said above, his 'axiom of solvability' is not valid. We now know that even completely rigorously stated problems, that is problems expressed in the completely formal language of logic, cannot, in general, be solved. That is, there are logical sentences whose validity cannot be determined by any general procedure or process. This contradicts Hilbert's statement in his lecture when he said of some of the great unsolved problems: 'However unapproachable these problems may seem to us and however helpless we stand before them, we have, nevertheless, the firm conviction that their solution must follow by a finite number of purely logical processes'. The problem of finding a method which tells whether a sentence is valid or not was formulated in *Grundzüge der Theoretischen Logik* (Hilbert and Ackermann, 1950) and is known as the *Entscheidungsproblem.*

The discovery that there is no such method was due to two people working independently (one in the USA, the other in England). The remarkable thing was, as often seems to happen with problems of this sort, that they both seemed to make the discovery, by different methods, at the same time, around 1936. These two formalisms are very different: one deals with functions defined by application and abstraction: the other deals with an abstract machine which was taken as the model for the first ever stored-program computer. Nevertheless they were both used to produce the same result.

3.3.1 Church and functions

The person working in the USA was Alonzo Church. He introduced what we are now familiar with as the **λ-calculus** (a good introduction is in Henson (1987)) as the vehicle for his proof that predicate calculus is undecidable. The λ-calculus has had a huge impact on computer science and we mention some of its effects here. Details of the proof are in Church (1936).

There has been a long history of using the ideas on abstraction and application that the λ-calculus formalizes as a basis for programming languages. The first well-known language with this basis was Lisp, as developed by John McCarthy (McCarthy, 1960). Later, in the 1960s, Peter Landin (Landin, 1965) initiated the idea of using λ-calculus as a basis for describing programming languages. This work was taken up by Strachey and his followers and developed into a method for giving denotational semantics for programming languages, and is well presented by Stoy (1977). It was also taken as a basis for new programming languages, which led to ISWIM (If You See What I Mean, Landin, 1966), which in turn has been further developed by others. The influence of ISWIM can be seen in languages such as SML (though SML was originally just ML and used for a particular purpose; for more on this see Chapter 9 on LCF) and the 'blackboard notation' initially developed by Landin and developed as a more complete programming notation in Bornat (1987).

3.3.2 Turing and mechanism

The person working in England was Alan Turing, and his method for the unsolvability of the *Entscheidungsproblem* involved the introduction of what we now know as **Turing machines**. A good introduction, which also gives a proof of the undecidability, is in Boolos and Jeffrey (1980). Turing's original paper (Turing, 1936) is very clear (though it contains mistakes in the detail which he later corrected). Turing's proof is based on showing that there are certain completely specifiable problems, such as 'does this computation ever print a 0?', which cannot be answered by algorithmic means, in other words by following a finite set of rules precise enough to be executed mechanically. He showed that if it were possible to decide validity in predicate logic then we would be able to answer questions such as the one above. This is clearly a contradictory state of affairs. Since the reasoning that leads to these conclusions is clearly sound the only way out of the contradiction is to deny the truth of the assumption that gave rise to it. Namely, we have to conclude that the validity problem is undecidable.

A direct link between λ-calculus and mechanism was provided in a paper of Landin's (Landin, 1964) which introduced the SECD machine (the store, environment, control, dump machine) and showed that the λ-calculus could be given a completely mechanical, that is operational, semantics which coincides exactly with the semantics given by Church, which was in terms of re-writing expressions. This work has also had a lasting impact in that it has formed the basis of work on implementation of the so-called functional languages, languages based on a 'sugaring' (to use a term of Landin's) of the somewhat terse language of the λ-calculus itself.

3.3.3 Computability

We have referred to the link between the λ-calculus and mechanism provided by the SECD machine. It was known long before 1964 that the λ-calculus and Turing machines pick out exactly the same class of functions, which are what we call the **computable** functions or recursive functions or effective functions. By 'effective functions' we mean 'the functions which it is in principle possible for someone to calculate'. Of course this is not a very exact definition, and it never can be since it is not clear how to define exactly a notion that depends on the possibilities of human calculation; to paraphrase Russell, we would need to formalize something which seems to be just a medical fact.

However, it is now commonly held to be the case that the class of functions definable by Turing machines or λ-calculus (or one of many other methods) does indeed coincide exactly with what is in principle calculable. That is, we have no evidence against this and much in favour of it. But, because we cannot formally define what we mean by terms such as 'calculable in principle' we cannot hope to prove a theorem. However, the correspondence is recorded as a thesis, a statement that is believed to be true. This is known as the **Church–Turing thesis**:

> The class of effectively calculable functions is the same as the class of Turing computable (or λ-definable or recursive) functions.

So, we have seen that there are problems which can be specified or described in completely formal detail, yet which no one has ever written or will ever write a program to solve. This is at once exciting, in that it gives us insight into a fundamental limit to computation, mysterious, in that it raises the question of whether human beings have the same kind of inherent limitation, and depressing since for the first time it seems that there is a definite bound to the tasks that computation can solve.

3.4 Models and theories

A sentence in a first-order language is just a collection of symbols put together according to certain syntactic rules. To give meaning to the sentences of a language we have first to identify the objects about which we are talking. We have called this collection of objects the universe, taken to be non-empty. Over the universe we have certain relations between objects. If our universe is a collection of people then such a relation might be 'p is the father of q'. We might take a particular person and identify that person's father. So, to say useful things about a universe we also want to be able to express relationships and generate new objects. This is where predicates and functions are used. We can express the relation above as $F(p, q)$ or, given a person q, we can identify the father by applying a function, $f(p)$, where f is the 'father producing' function. This is the idea behind interpretations as defined above.

So, the interpretations give a basis upon which to judge the truth or falsity of a sentence in a particular language. In a given interpretation each predicate has a truth-value when its parameters are replaced by names. Such sentences are called **ground atoms**. Each ground atom has a value of true or false, that is if L is a ground atom then either $\neg L$ is true or L is true. The truth-values calculated from the ground atoms are then combined by the connectives according to their usual meanings to give finally a truth-value for the whole sentence. From these ideas we can define the notion of a model.

Definition 3.6: A **model** is an interpretation of a sentence which makes the sentence true. We also extend this to sets of sentences, G, and say that M is a model of G iff M is a model for every sentence in G.

Definition 3.7: A **theory** is a set of sentences in a language which are true in a class of interpretations.

A set G of sentences determines a class of interpretations I, that is all those interpretations that are models of G, and I determines a further set of sentences that are true in all the interpretations in I. We write the set of sentences that are true in all models of G as $\mathrm{Th}(G)$, that is the theory based on G.

Definition 3.8: We call a set such as G the **axioms** (of the theory). The sentences which are elements of the theory we call the **theorems**.

If $\mathcal{S}$ is a theorem of the theory determined by the set of axioms G then we write $G \models \mathcal{S}$, which we have introduced already as entailment, since here theoremhood is determined semantically.

Theories provide the starting point for the use of logic in formalizing reasoning in other areas of mathematics and computer science, and we shall be giving some examples in Chapter 6. To apply logic, however, you have to be able to prove theorems. We have already seen one method of doing this with axiom systems. In the next two chapters we look at some other, more powerful, proof methods.

SUMMARY

- As a means of formalizing arguments, propositional calculus lacks expressive power because there is no way of referring to individuals or stating that propositions apply to some or all individuals. Predicate calculus is an extension of propositional calculus with names, variables, function symbols, predicate symbols and quantifiers.
- In predicate calculus individuals are denoted by *terms*. A term is either a name, a variable symbol or a function symbol applied to terms. The propositional variables of the propositional calculus are extended to *atomic formulas* which have the form of predicate symbols applied to terms. The quantifiers *bind* variables in the formulas over which their *scope* extends. The notions of scope and *free* and *bound variable* are similar to those in computer science. The other logical connectives are the same as in propositional logic and the recursive definition of compound formulas is similar.
- The simple truth-tables of propositional calculus no longer suffice to give a semantics for predicate calculus. The notion of *interpretation* is introduced. An interpretation consists of a universe of objects, an assignment function that specifies which objects in the universe the terms of the language denote, and a function that maps predicate symbols in the language to relations over the universe. The denotation of a term (the element of the universe that it denotes) is calculated by means of the recursively defined assignment function which maps names and variable symbols to elements of the universe, and function symbols to functions over the universe.
- For each of the logical connectives and quantifiers, rules are given specifying when a well-formed formula in which they are the principal connective is *satisfied* for a given assignment

in a given interpretation. Atomic formulas are satisfied if the denotation of their arguments is in the relation denoted by their predicate symbol. A formula of predicate calculus is *true* in an interpretation if it is satisfied for all assignments in that interpretation, and it is *valid* if it is true in all interpretations.

- A formal deductive system for predicate calculus is given as an extension of that for propositional calculus. The notion of proof as a sequence of formulas in which each element is either an axiom, or is derived from earlier members of the sequence by rules of inference, is also similar.
- Whereas propositional calculus is *decidable* – any sentence can be shown either valid or not valid in a finite number of steps by the method of truth-tables – this is not the case for the full predicate calculus. There can be no mechanical method for deciding that an arbitrary formula of predicate calculus is not valid. This fundamental result was proved in two different ways, by Church who invented the λ-calculus to do it, and by Turing who modelled the notion of computability with an idealized machine that turned out to be the forerunner of the stored-program computer as we know it today.
- A *theory* is a set of sentences all of which are true in a class of interpretations. A *model* is an interpretation that makes all the sentences of a theory true. Theories and models provide the starting point for the use of logic in formalizing areas of mathematics such as arithmetic and set theory, and data abstractions in computer science such as lists and trees (see Chapter 6).

Semantic Tableaux

4.1 Introduction

In this chapter we introduce one of the main methods that we shall be using in the rest of the book to judge validity and consistency.

Semantic tableaux will turn out to be much more expressive and easy to use than truth-tables, though that is not the reason for their introduction.

In any model for G, a set of sentences, each element $\mathscr{g}_i$ of G is true. Thus the conjunction $\wedge_i \mathscr{g}_i$ is also true. Our problem, in proving that $G \models \mathscr{S}$ for some sentence $\mathscr{S}$, is to show that if $\wedge_i \mathscr{g}_i$ is true then $\mathscr{S}$ is true whatever model of G we choose. That is, we must show that

$$(\wedge_i \mathscr{g}_i \rightarrow \mathscr{S}) \quad \textbf{(4.1)}$$

is valid.

We can think of this as working by saying that the 'if $\wedge_i \mathscr{g}_i \ldots$' picks out all the models from all the interpretations of its language and then passes these on to the '... then $\mathscr{S}$'.

Because of the definition of implication, if a given interpretation is not a model of $\bigwedge_i \mathcal{G}_i$ then this will be false and so the whole of the relation 4.1 is true; and if the given interpretation of $\bigwedge_i \mathcal{G}_i$ is a model then the truth of 4.1 depends on the truth of $\mathcal{S}$. Thus the relation 4.1 expresses the statement that $\mathcal{S}$ is a theorem of the theory generated by G, that is $G \models \mathcal{S}$.

As it happens the relation 4.1 does not express this in quite the way that we wish. However, consider the following argument. If $(\bigwedge_i \mathcal{G}_i \to \mathcal{S})$ is valid then for each interpretation in its language

(1) if $\bigwedge_i \mathcal{G}_i$ is true then $\mathcal{S}$ must be true (by the definition of implication) so $(\bigwedge_i \mathcal{G}_i \wedge \neg \mathcal{S})$ must be false for every such interpretation;

(2) if $\bigwedge_i \mathcal{G}_i$ is false then the value of $\mathcal{S}$, true or false, has no effect and $(\bigwedge_i \mathcal{G}_i \wedge \neg \mathcal{S})$ is false for every such interpretation;

(3) for each interpretation only (1) or (2) holds, since $\bigwedge_i \mathcal{G}_i$ can only be true or false, and thus $(\bigwedge_i \mathcal{G}_i \wedge \neg \mathcal{S})$ is unsatisfiable.

This final sentence is the form that is most suitable for our purpose. $\mathcal{S}$ is a theorem of the theory generated by G if and only if $(\bigwedge_i \mathcal{G}_i \wedge \neg \mathcal{S})$ is unsatisfiable.

This argument forms the basis of the method of semantic tableaux. Given that we are required to prove $G \models \mathcal{S}$ then we must attempt to show that $(\bigwedge_i \mathcal{G}_i \wedge \neg \mathcal{S})$ is unsatisfiable. The basic idea is to allow this sentence to be elaborated in every possible way that expresses its satisfiability. If we find that no such way exists then we conclude that it is unsatisfiable, that is there is no assignment which will allow it to be true. Then we know that $G \models \mathcal{S}$ from the preceding argument.

Of course, this is just another way of saying that $G \models \mathcal{S}$ iff $G \cup \{\neg \mathcal{S}\} \models$, which we have seen before in Chapter 2.

4.2 Semantic tableaux for propositional calculus

In this section we shall be dealing only with a propositional language. As we saw in Chapter 2 for such a language the notion of interpretation becomes trivial. It is simply a mapping from propositional letters to the truth-values t and f, that is it is what we called a valuation before.

Given a set of sentences of the form $G \cup \{\neg \mathcal{S}\}$ we want to show that no model for it exists, that is no valuation satisfies it. We say that the semantic tableau method is a 'proof by refutation' method. We want to refute the existence of a model of $G \cup \{\neg \mathcal{S}\}$.

We display the evolution of the refutation in the form of a binary tree, using the terms **root**, **leaf** and **path** with their usual meanings.

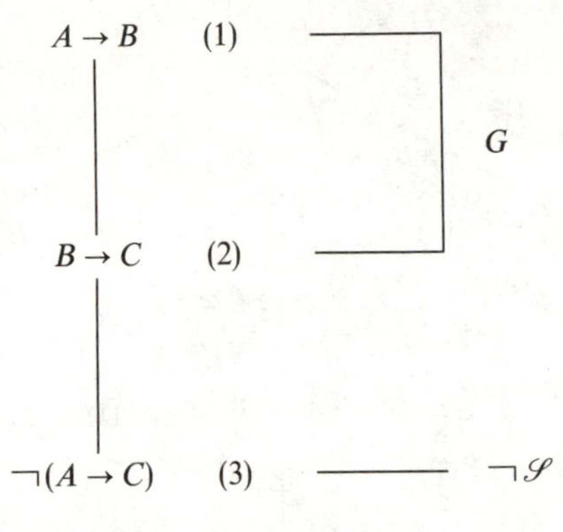

Figure 4.1.

Definition 4.1: We have that a **binary tree** is given by a pair $B = \langle s, N \rangle$ where s is a **successor function** and N is a set of **nodes**. Given an $n \in N$, $s(n) \subseteq N$ is the **successor set** of N, that is the set of nodes which come 'below' n in the tree. Since B is binary $s(n)$, for any n, has at most two members. Since B is a tree any node appears in at most one successor set. The **root** node of B is the unique node which is in the successor set of no node. A **leaf** is a node which has an empty successor set. A **path** is a sequence of nodes $\langle n_i \rangle \{i \geqslant 0\}$ such that $n_{i+1} \in s(n_i)$, $i \geqslant 0$.

We also have a labelling function which maps nodes to their labels, which will be sentences. Consider the set of sentences shown in Figure 4.1. Here we have a tree with the node set $\{n_0, n_1, n_2\}$ and $s(n_0) = \{n_1\}$, $s(n_1) = \{n_2\}$ and $s(n_2) = \{\ \}$. The labelling function l is given by $l(n_0) = A \rightarrow B$, $l(n_1) = B \rightarrow C$ and $l(n_2) = \neg(\mathrm{A} \rightarrow \mathrm{C})$. Note here that our set of propositional letters has $\{A, B, C\}$ as a subset.

Sentences listed vertically beneath one another are, by definition, understood to be conjugated together.

It is easily seen that $G \vDash \mathscr{S}$ so how do we use the tableau method to prove it? Well, we must show that the above set is unsatisfiable, that is that it has no model.

Definition 4.2: The sentences in $G \cup \{\neg\mathscr{S}\}$ are the **initial sentences** of the tableau. $G \vDash \mathscr{S}$ is the **entailment represented by** the initial sentences given by $G \cup \{\neg\mathscr{S}\}$.

Referring to Figure 4.1, we take sentence (1) first. According to what we said before, the sentence must be allowed to express its validity in every possible way. We can see the possible ways that (1) can be valid by looking at

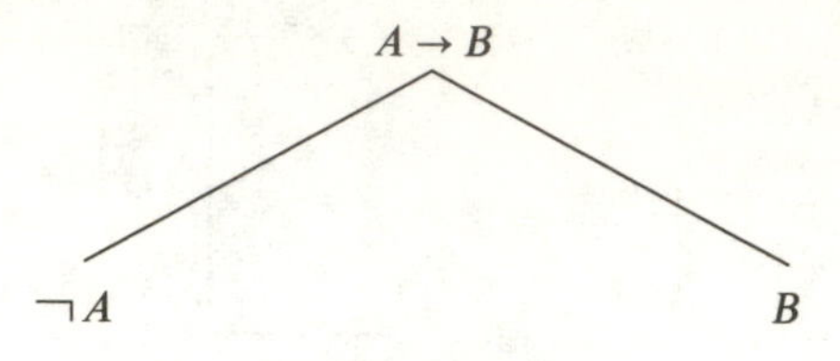

Figure 4.2.

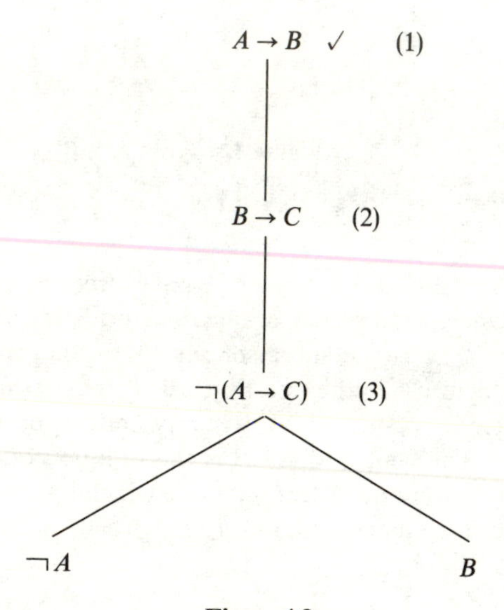

Figure 4.3.

the truth-table for the connective →

→	t	f
t	t	f
f	t	t

If we take the cases where the antecedent A is false together with those where the consequent B is true then we have the cases which are exactly those which make $A \to B$ true. The fact that there is some redundancy here is not important. All that we require is that the tableau expresses all the possible ways that $A \to B$ can be true. Referring now to Figure 4.2, the fork in the tableau denotes a disjunction between A and B. $A \to B$ is true iff A is false or B is true (or both).

The tableau is now as shown in Figure 4.3 and we mark sentence (1) with a ✓ to show that all its cases have been covered already. We use the same argument with sentence (2) to get the tableau shown in Figure 4.4.

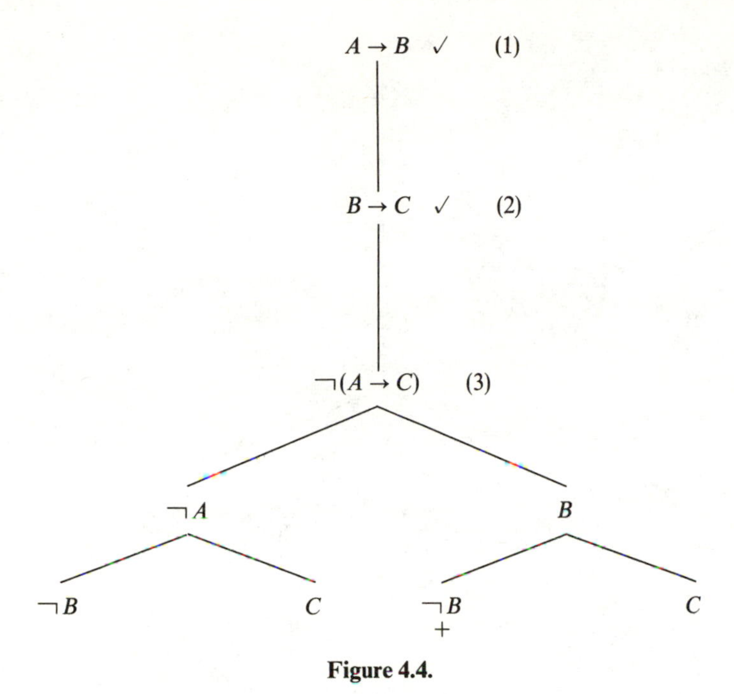

Figure 4.4.

Note that the path with + at its leaf contains the sentences $A \to B$, $B \to C$, $\neg(A \to C)$, B and $\neg B$. The last two form a contradiction so we close the path, that is we mark it with + to show that it should take no further part in the refutation. There is no possible interpretation that could allow this path to be satisfiable.

Definition 4.3: We say that a path is **closed** iff it contains a sentence and its negation. A tableau is **closed** iff all its paths are closed.

We have now to deal with sentence (3). For $\neg(A \to C)$ to be true $A \to C$ must be false and by referring to the above truth-table (with B as C) we see that to make this so A must be true and B false so we add a new path to the still open paths of the tableau to get the tableau in Figure 4.5, so the whole tableau is now closed.

We can now say that our initial entailment $G \models \mathscr{S}$ is valid. This is because the methods used to expand the sentences covered all the possible ways in which each could be true. However, we have seen that every path is unsatisfiable so there is no way for all the initial sentences to be true, that is they form an unsatisfiable set. So no model exists for $G \cup \{\neg\mathscr{S}\}$.

If we use this method on an entailment $G \models \mathscr{S}$ which is not valid then we will reach a stage when no more sentences can be expanded but the tableau is unclosed. In this case each unclosed path forms a model which satisfies $G \cup \{\neg\mathscr{S}\}$ since, by the method of construction, we have that the

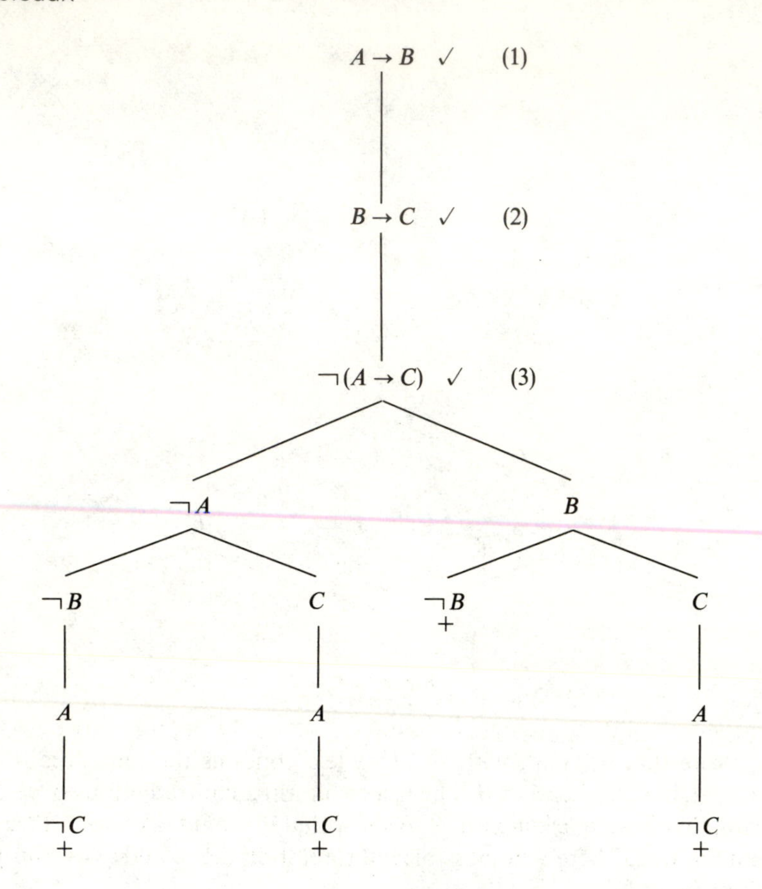

Figure 4.5.

Figure 4.6.

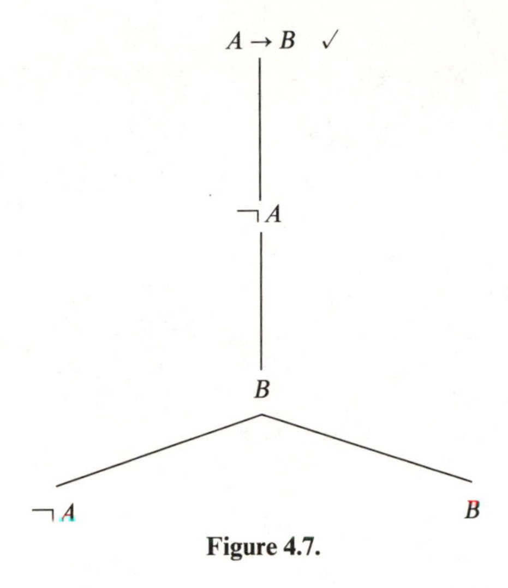

Figure 4.7.

sentences on an open path form a satisfiable set of sentences. Hence, there is a valuation under which all the sentences on an open path are true. It follows that, since the initial sentences are on every path, there is some valuation under which they are all true. The simplest such valuation is one where each sentence that appears negated has the value f and any others, that is the unnegated ones, have value t. For example the initial tableau in Figure 4.6 produced the completed tableau in Figure 4.7, which says that a valuation v such that $v(A) = \text{f}$ and $v(B) = \text{t}$ satisfies the initial sentences, as you can easily check. So, $G \cup \{\neg \mathscr{S}\}$ is satisfiable so $G \vDash \mathscr{S}$ is an invalid entailment, that is $\{A \to B, \neg A\}$ does not entail $\neg B$.

You should compare this with the truth-table that we would get for this entailment:

A	B	$\neg A$	$\neg B$	$A \to B$
t	t	f	f	t
t	f	f	t	f
f	t	t	f	t
f	f	t	t	t

and here we can see that in the third line we have the premises true while the conclusion is false. That is, the entailment is invalid, since the conclusion is not true everywhere that the premises are, and on this line we see that A has value f and B has value t, just as with the tableau above.

The method given here can be extended to handle the other connectives and the relevant direct extensions are shown in Figure 4.8. We call these the (schemas for) splitting rules (for connectives) of the tableau

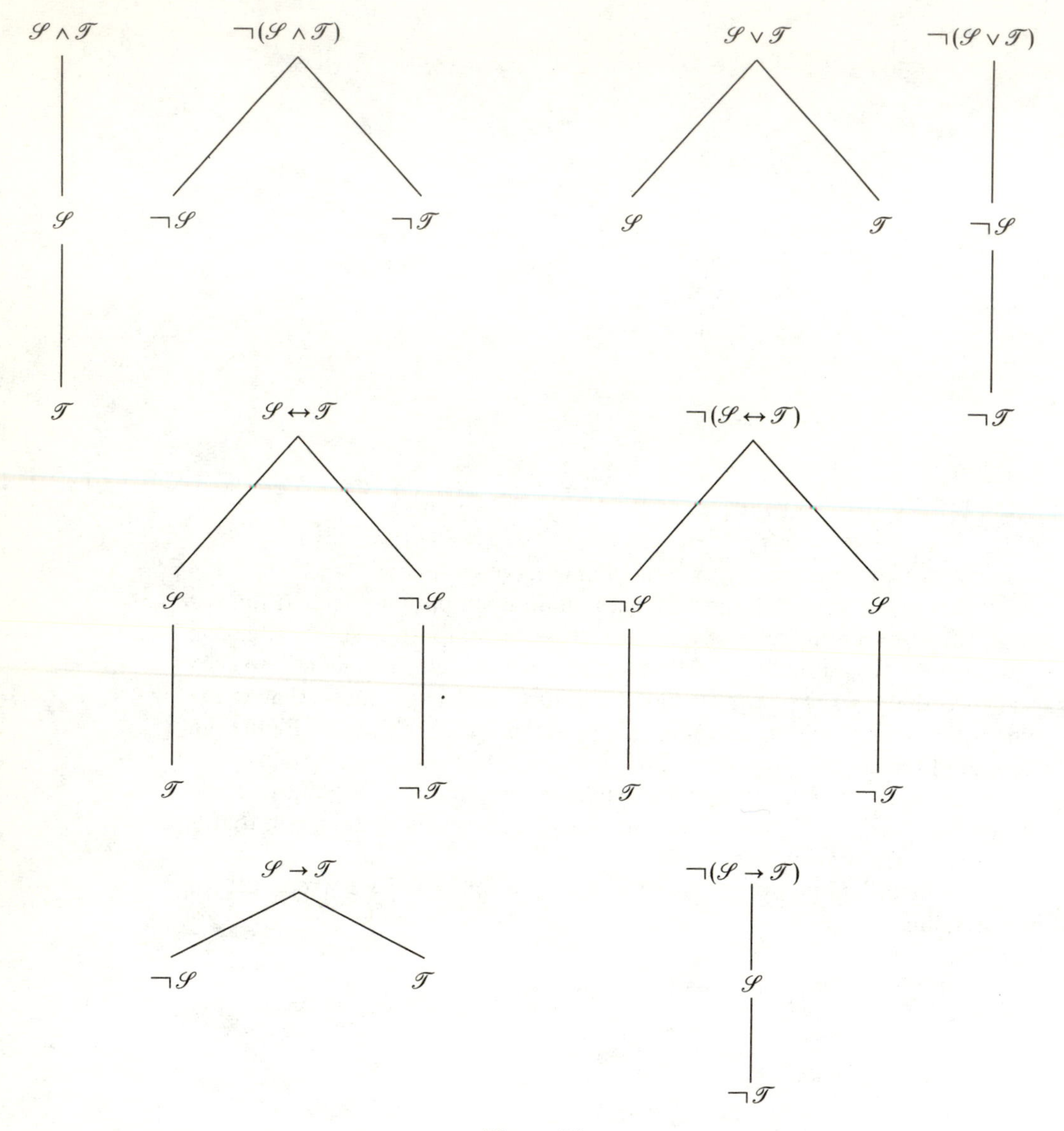

Figure 4.8.

method. They may be verified as for $\rightarrow$ by comparing with the usual truth tables for the connectives.

As a final example we give a proof of

$$\{(A \wedge B) \rightarrow C, \neg A \rightarrow D\} \vDash B \rightarrow (C \vee D)$$

the tableau for which is given in Figure 4.9.

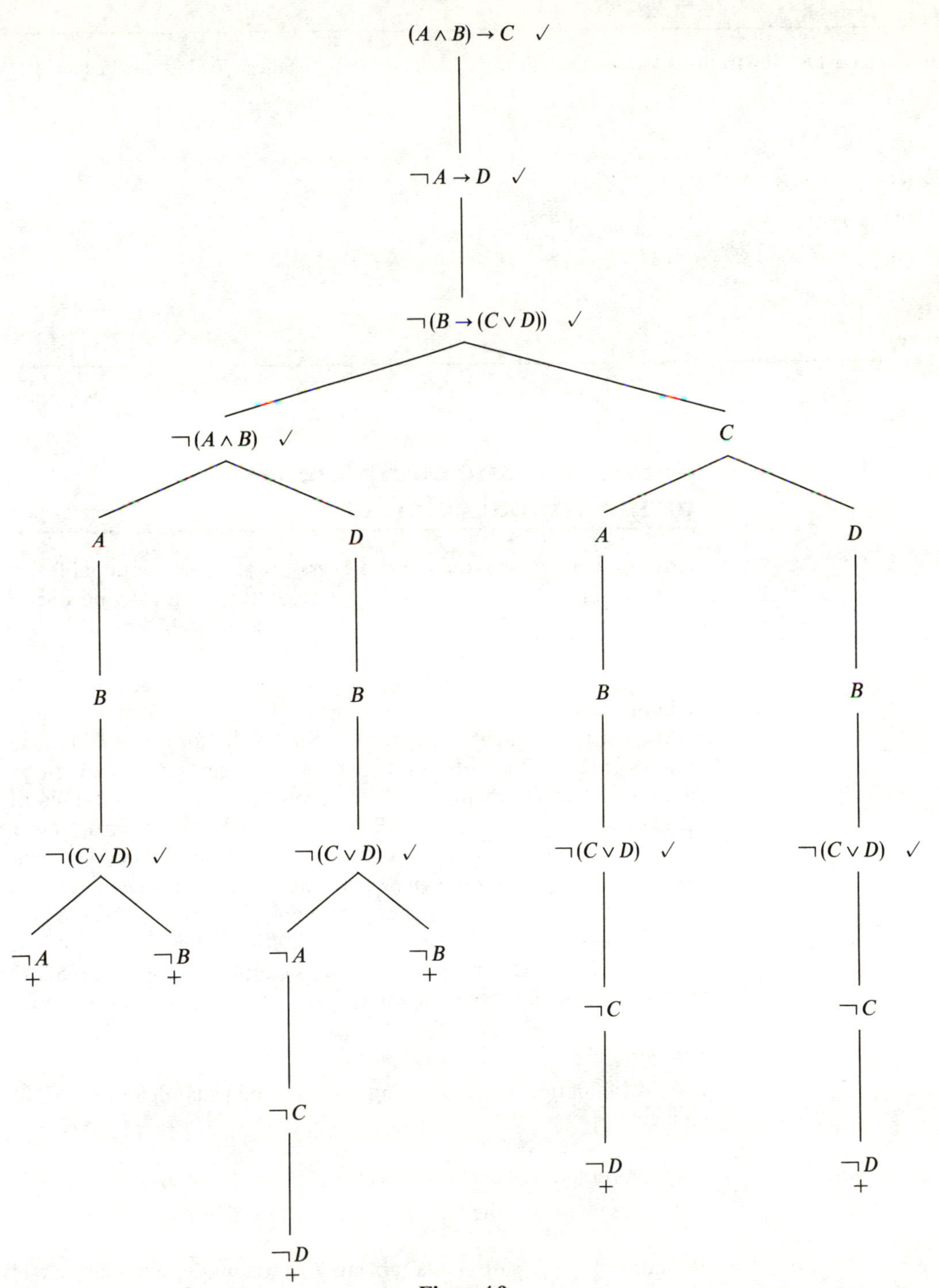

Figure 4.9.

Exercise 4.1 Use truth-tables and tableaux to decide on the status of the following entailments:

(a) $\models A \leftrightarrow (A \wedge B)$

(b) $\models (A \vee B) \leftrightarrow (B \vee A)$

(c) $\{A \leftrightarrow B, A \vee B\} \models A \wedge B$

(d) $\{A\} \models A \vee B$

(e) $\{A, B\} \models A \wedge B$

(f) $\{A \wedge B\} \models A$

(g) $\{A \vee B, A \vee (B \wedge C)\} \models A \wedge C$

4.3 Soundness and completeness for propositional calculus

We can now do a soundness and completeness proof for semantic tableaux. That is, we can prove, as we did for the formal system for propositional calculus in Chapter 2, that a sentence is provable by tableaux iff it is a tautology.

Definition 4.4: A sentence S of the form $A \wedge B$, $\neg(A \vee B)$ or $\neg(A \rightarrow B)$ will be called a **sentence of type α**. If S is of the form $A \wedge B$ then $\alpha 1$ denotes A and $\alpha 2$ denotes B. If S is of the form $\neg(A \vee B)$ then $\alpha 1$ denotes $\neg A$ and $\alpha 2$ denotes $\neg B$. If S is of the form $\neg(A \rightarrow B)$ then $\alpha 1$ denotes A and $\alpha 2$ denotes $\neg B$. A sentence S of the form $A \vee B$, $\neg(A \wedge B)$, $A \leftrightarrow B$, $A \rightarrow B$ or $\neg(A \leftrightarrow B)$ will be called a **sentence of type β**. If S is of the form $A \vee B$ then $\beta 1$ denotes A and $\beta 2$ denotes B. If S is of the form $\neg(A \wedge B)$ then $\beta 1$ denotes $\neg A$ and $\beta 2$ denotes $\neg B$. If S is of the form $A \rightarrow B$ then $\beta 1$ denotes $\neg A$ and $\beta 2$ denotes B. If S is of the form $A \leftrightarrow B$ then $\beta 1$ denotes $A \wedge B$ and $\beta 2$ denotes $\neg A \wedge \neg B$. If S is of the form $\neg(A \leftrightarrow B)$ then $\beta 1$ denotes $\neg A \wedge B$ and $\beta 2$ denotes $A \wedge \neg B$.

Lemma

For any valuation v the following hold, by the usual definition of the connectives:

(C1) A sentence of the form α is true iff both $\alpha 1$ and $\alpha 2$ are true.

(C2) A sentence of the form β is true iff $\beta 1$ or $\beta 2$ is true.

Definition 4.5: A path p in a tableau T is **true** under a valuation v iff all the sentences on p are true under v. T is **true** iff some paths are true.

Definition 4.6: A tableau T_2 is a **direct extension** of a tableau T_1 if it can be obtained from T_1 by application of one of the following rules, where p_1 is a path in T_1:

(D_1) If some α occurs on path p_1 with leaf l, then join $\alpha 1$ to l and $\alpha 2$ to $\alpha 1$ to get T_2.

(D_2) If some β occurs on path p_1 with leaf l then add β_1 as the left successor of l and $\beta 2$ as the right successor of l to get T_2.

Lemma

For any satisfiable set G of sentences the following hold:

(i) (S_1) If $\alpha \in G$ then $G \cup \{\alpha 1, \alpha 2\}$ is satisfiable.
(S_2) If $\beta \in G$ then $G \cup \{\beta 1\}$ or $G \cup \{\beta 2\}$ is satisfiable.

(ii) If a tableau T_2 is a direct extension of a tableau T_1 which is true under v then T_2 is true under v.

Proof

(i) S_1 and S_2 follow from C_1 and C_2.

(ii) If T_1 is true under v then it contains at least one true path, say p, on which the set of sentences G appears, all members of which are true under v. T_2 is a direct extension of T_1, so it was obtained by one of the steps D_1 or D_2 above applied to a path p' of T_1. If p' is different from p then p is still a path of T_2 and hence T_2 is true under v.

If p' is the same as p then we have two cases which are proved by use of S_1 and S_2.

For instance, if p was extended by D_1 then some α appeared on p so by S_1 the sentences in $G \cup \{\alpha 1, \alpha 2\}$ are true under v and appear on p', so p' is true too. The proof of the other case is similar. ■

Corollary

If the initial sentences of a tableau T generated by D_1 and D_2 are satisfied then T must be true.

Definition 4.7: $G \models \mathscr{S}$ is **provable** iff there is a tableau with initial sentences $G \cup \{\neg \mathscr{S}\}$ which has an extension which is closed.

Soundness theorem

Any entailment provable by a tableau must be valid.

Proof A closed tableau cannot be true under any valuation. Hence, by the corollary, the initial sentences cannot be satisfiable. Thus, the entailment represented by the initial sentences is valid. ■

Definition 4.8: G is a **model set** in the following cases:

(H_0) No proposition symbol and its negation are both in G.
(H_1) If $\alpha \in G$, then $\alpha 1, \alpha 2 \in G$.
(H_2) If $\beta \in G$, then $\beta 1 \in G$ or $\beta 2 \in G$ or both.

Lemma

If G is a model set then G is satisfiable.

Proof We prove this by constructing a model for G. Let v be a valuation of the sentences in G. Let q be a propositional letter that occurs in at least one element of G. Then if $q \in G$ we put $v(q) = \mathrm{t}$, if $\neg q \in G$ then $v(q) = \mathrm{f}$ and if neither then we arbitrarily give $v(q)$ the value t.

Now, take an arbitrary element X of G. Let X have $n > 0$ connective symbols in it. Suppose that all elements of G with $m < n$ connectives in them are true under v. Then X is true under v also. We show this by induction. If X is an α then $\alpha 1$ and $\alpha 2$ must also be in G, by H_1. $\alpha 1$ and $\alpha 2$ both contain less than n connectives and so are true. Thus, X is true. The case for X being a β is similar.

It is clear, then, that we have constructed a model for G, so G is satisfiable. ■

Definition 4.9: A path p in a tableau is **complete** iff

(i) for every α on p, both $\alpha 1$ and $\alpha 2$ are on p;
(ii) for every β on p, at least one of $\beta 1$ or $\beta 2$ is on p.

Definition 4.10: A tableau is **completed** iff every path is either closed or complete.

We can describe one particular way of repeatedly directly extending a tableau so as to give a sound and complete proof method as follows:

```
begin
  repeat
    changes:=false
    close each path that contains a sentence and its negation
    if all paths are closed
    then deliver "correct entailment"
    else if a connective rule can be applied
        then
        changes := true
        apply the appropriate rule
        mark the sentence as used
```

```
    until not changes
    deliver "incorrect entailment"
end
```

This algorithm can be seen to achieve the required soundness and completeness conditions by noticing that at the beginning of each time around the loop the tableau is open and changes is true. At the end of the loop, (i) all the paths are closed and we deliver correct entailment or (ii) a splitting rule has been applied (and at least one path is open) or (iii) no splitting rule is applicable (and at least one path is open).

In case (i) we are obviously finished since the tableau is closed and by the soundness theorem the entailment must be valid. In case (ii) we added a finite number of sentences to the tableau, each of which had one less connective than the split sentence that produced them. In case (iii) we have again finished, but this time by the completeness theorem we do not have a valid entailment.

Thus, we only re-enter the loop in case (ii). Clearly, the number of occurrences of connectives in unmarked sentences is strictly decreasing so we will eventually reach a stage where no connectives appear in unmarked sentences, and thus the algorithm always terminates. We expected this since the propositional calculus is decidable.

Notice that this is only one possible algorithm since the theory does not specify an order for direct extensions, so we can change the algorithm by imposing different orders on the selection of sentences to be split. Decisions about which ordering, if any, to impose on the use of rules will be guided by how efficient we want the procedure to be. That is, how much work it takes actually to decide on the status of a given sentence. In general we can show that any procedure will be very inefficient on many sentences and the development of more or less efficient procedures for deciding logical status comes under the remit of people doing research in automated reasoning. We will not go into these matters further here, but suffice it to say that there is a huge body of research on this topic which is still growing very rapidly.

The algorithm above was written in a sort of procedural English, and is sometimes called pseudo-code. This is done to suggest how such an algorithm might be written in your favoured programming language. Below we give SML and Prolog versions of, essentially, this algorithm so that you can compare how these languages would go about implementing it.

The first language we use is SML. The following first collection of functions carries out the operations on sentences that are needed. In particular, apply_rule takes a sentence and returns its subsentences as used by the connective rules:

```
fun Neg (Not(sent)) = sent
  | Neg sent = Not(sent);
```

```
fun alpha(And(s,t)) = true
  | alpha(Not(Imp(s,t))) = true
  | alpha(Not(Or(s,t))) = true
  | alpha(s) = false;

fun composite(Var s) = false
  | composite(Not(Var s)) = false
  | composite(s) = true;

fun apply_rule (Imp(s,t)) = (Neg(s),t)
  | apply_rule (Eq(s,t)) = (And(s,t),And(Neg(s),Neg(t)))
  | apply_rule(And(s,t)) = (s,t)
  | apply_rule(Or(s,t)) = (s,t)
  | apply_rule(Not(Imp(s,t))) = (s,Neg(t))
  | apply_rule(Not(Eq(s,t))) = (And(Neg(s),t),And(s,Neg(t)))
  | apply_rule((Not(And(s,t))) = (Neg(s),Neg(t))
  | apply_rule(Not(Or(s,t))) = (Neg(s),Neg(t));
```

Now we need to introduce a type that will model the tableau. We define three kinds of node: a Leaf; a Onenode, that is a node with one successor; a Twonode, that is a node with two successors. Then, as we will see later on, we can use these three kinds of nodes to describe what action to do on any tableau. Each node, of whichever kind, also contains information to say whether it has been Closed or Used:

```
type nodeinfo = {Closed : bool, Used : bool};
datatype Tableau = Leaf of (SENT * nodeinfo) |
                   Onenode of (SENT * nodeinfo * Tableau) |
                   Twonode of (SENT * nodeinfo * Tableau * Tableau);
```

These next functions do the building up of the tableau; initinfo builds the first part of the tableau from the initial sentences:

```
val initinfo = {Closed = false,Used = false};

fun initialize (nil,goal) = Leaf(Neg(goal,initinfo))
  | initialize (premise::nil,Null) = Leaf(premise,initinfo)
  | initialize (premise::rest,goal) = Onenode(premise,initinfo,initialize(rest,goal));

val useup = {Closed = false,Used = true};
val useable = (fn a => fn b : nodeinfo => not (#Used(b)) andalso composite(a));

fun extend(Leaf(S,N),sent) =
          if not(#Closed(N))
          then let val (f,s) = apply_rule(sent)in
                   if alpha(sent)
                   then Onenode(S,N,Onenode(f,initinfo,Leaf(s,initinfo)))
                   else Twonode(S,N,next(Leaf(f,initinfo)),next(Leaf(s,initinfo)))
               end
          else (Leaf(S,N))
```

```
  | extend(Onenode(S,N,T),sent) =
                                  if not (#Closed(N))
                                  then Onenode(S,N,extend(T,sent))
                                  else Onenode(S,N,T)
  | extend(Twonode(S,N,T1,T2),sent) =
                                   if not (#Closed(N))
                                   then Twonode(S,N,extend(T1,sent),extend(T2,sent))
                                   else Twonode(S,N,T1,T2)
and next(Leaf(S,N)) =
                         if (useable S N)
                         then extend(Leaf(S,useup),S)
                         else (Leaf(S,N))
  | next(Onenode(S,N,T)) =
                              if (useable S N)
                              then extend(Onenode(S,useup,T),S)
                              else (Onenode(S,N,next T))
  | next(Twonode(S,N,T1,T2)) =
                                if (useable S N)
                                then extend(Twonode(S,useup,T1,T2),S)
                                else (Twonode(S,N,next T1,next T2));
```

The next functions are used to see whether or not the tableau is closed:

```
val closeit = fn N : nodeinfo => {Closed = true,Used = #Used(N)};

fun cleaves(Leaf(S,N)) = Leaf(S,closeit N)
  | cleaves(Onenode(S,N,T)) = Onenode(S,closeit N,cleaves(T))
  | cleaves(Twonode(S,N,T1,T2)) = Twonode(S,closeit N,cleaves(T1),cleaves(T2));

fun close'(S,Leaf(S',N)) = if S = S'
                           then cleaves(Leaf(S',N))
                           else Leaf(S',N)
  | close'(S,Onenode(S',N,T)) =
                               if S = S'
                               then cleaves(Onenode(S',N,T))
                               else Onenode(S',N,close'(S,T))
  | close'(S,Twonode(S',N,T1,T2)) =
                                if S = S'
                                then cleaves(Twonode(S,N,T1,T2))
                                else Twonode(S',N,close'(S,T1),close'(S,T2));

fun Close (Leaf(S,N)) = close'(Neg(S),Leaf(S,N))
  | Close(Onenode(S,N,T)) = Onenode(S,N,close'(Neg(S),Close(T)))
  | Close(Twonode(S,N,T1,T2)) = Twonode(S,N,close'(Neg(S),Close(T1)),
                                              close'(Neg(S),Close(T2)));

fun closed (Leaf(S,N)) = #Closed(N)
  | closed (Onenode(S,N,T)) = #Closed(N) orelse closed T
  | closed (Twonode(S,N,T1,T2)) = #Closed(N) orelse
                                          ((closed T1) andalso
                                             (closed T2));
```

Finally, we can write a function for the main algorithm. Note that here instead of using a value 'changes' as we did in the pseudo-code above here we test for any changes by seeing whether or not the tableau is unchanged by comparing it with the result of applying another round of direct extensions, via next, to itself. This is, of course, very inefficient. However, what we have been trying to do here is to present the clearest SML version of something close to the algorithm above. There are more efficient versions given in Appendix B. The reader will see that, as so often, we trade simplicity of algorithm for efficiency of computation.

```
fun make_tableau T = let val T' = next T in
                         if T' = T
                         then T
                         else make_tableau(Close T')
                     end;

infix entails;
fun asslist entails goal =
          let val start_tableau = initialize(asslist,goal) in
             let val final_tableau = make_tableau(start_tableau) in
                if closed(final_tableau)
                then output std_out "\ncorrect\n"
                else output std_out "\nincorrect\n"
             end
          end;
```

Next, we give the Prolog program for the same algorithm.

```
:- op(500,xfx,    [:]).
:- op(510, fy,   [~]).
:- op(520,xfy,   [/\]).
:- op(530,xfy,   [\/]).
:- op(540,xfx, [->]).
:- op(550,xfx,    [?]).

Assumptions?Goal :- set_up(Assumptions,Goal,[],open,Start_tree,R),!,
   ((R = clsd,!,write('valid'));(maketab(Start_tree,[],Final_tree),
   ((closed(Final_tree),!,write('valid'));write('not valid')))).

set_up([],G,B,X,tr(NotG:U,Y),R) :- literal(~G,NotG:U),
   (((X = clsd;closes(NotG,B)),Y = clsd);Y = open),R = Y.
set_up([H|S],G,B,X,tr(A:U,[T]),R) :- literal(H,A:U),
   (((X = clsd;closes(A,B)),Y = clsd);Y = open),set_up(S,G,[A|B],Y,T,R).

maketab(tr(X,clsd),_,tr(X,clsd)).
maketab(tr(F:Used,open),B,tr(F:yes,T)) :-
   (Used = yes,T = open);(Used = no,apply_rule(F,[F|B],T)).
maketab(tr(F:Used,[T]),B,tr(F:yes,[NewT])) :-
   (Used = no,extend(T,B,F,Ta),maketab(Ta,[F|B],NewT));
   (Used = yes,                 maketab(T ,[F|B],NewT)).
```

```
maketab(tr(F:Used,[L,R]),B,tr(F:yes,[NewL,NewR])):-
   (Used=no,extend(L,B,F,La),extend(R,B,F,Ra),
                maketab(La,[F|B],NewL),maketab(Ra,[F|B],NewR));
   (Used=yes,maketab(L ,[F|B],NewL),maketab(R ,[F|B],NewR)).

extend(tr(F:U,clsd),_,_,tr(F:U,clsd)).
extend(tr(F:U,open),B,X,tr(F:U,T)):- apply_rule(X,[F|B],T).
extend(tr(F:U,[T]),B,X,tr(F:U,[NewT])):- extend(T,[F|B],X,NewT).
extend(tr(F:U,[L,R]),B,X,tr(F:U,[NewL,NewR])):-
   extend(L,[F|B],X,NewL),extend(R,[F|B],X,NewR).

apply_rule(    A/\B    ,Branch,Newtree):- alpha(  A,   B,Branch,Newtree).
apply_rule(    A\/B    ,Branch,Newtree):-  beta(  A,   B,Branch,Newtree).
apply_rule(    A -> B  ,Branch,Newtree):-  beta(~A,   B,Branch,Newtree).
apply_rule(~(A/\B)     ,Branch,Newtree):-  beta(~A,~B,Branch,Newtree).
apply_rule(~(A\/B)     ,Branch,Newtree):- alpha(~A,~B,Branch,Newtree).
apply_rule(~(A -> B),Branch,Newtree):- alpha(  A,~B,Branch,Newtree).

alpha(U,V,B,[tr(X:G,[tr(Y:H,T)])]):- literal(U,X:G),literal(V,Y:H),
   (((closes(X,B),!;closes(Y,[X|B])),!,T=clsd);T=open).

beta(U,V,B,[tr(X:G,L),tr(Y:H,R)]):- literal(U,X:G),literal(V,Y:H),
   ((closes(X,B),!,L=clsd);L=open),((closes(Y,B),!,R=clsd);R=open).

literal(X,Y:U):-
   dblneg(X,Y),(((atomic(Y);(Y= ~Z,atomic(Z))),U=yes);U=no).

dblneg(X,Y):- (X= ~(~Z),!,dblneg(Z,Y));Y=X.

closes(F,[H|T]):- F= ~H,!; ~F=H,!;closes(F,T).

closed(tr(_:_,T)):-
   T=clsd;(T=[X],closed(X));(T=[L,R],closed(L),closed(R)).
```

Completeness theorem

If $G \models \mathscr{S}$ is valid then it is provable.

Proof Assume that $G \models \mathscr{S}$ is valid. Then $G \cup \{\neg\mathscr{S}\}$ is unsatisfiable. Let $C(G, \mathscr{S})$ be a tableau produced by the algorithm above with initial sentences $G \cup \{\neg\mathscr{S}\}$. If some path p of $C(G, \mathscr{S})$ is unclosed then, since it is complete, the sentences on p form a model set and thus are satisfiable.

Hence $G \cup \{\neg\mathscr{S}\}$ is satisfiable, so it is not the case that $G \models \mathscr{S}$. Thus, no path of $C(G, \mathscr{S})$ can be unclosed. Thus $G \models \mathscr{S}$ is provable. ■

Since if every path of $C(G, \mathscr{S})$ above must close then every path must be of finite length and $C(G, \mathscr{S})$ is finitary, it follows that it will close after some finite number of steps. Thus we have

Corollary

If $G \models \mathscr{S}$ is valid then it is provable in a finite number of steps.

From the above it is clear that we have

Correctness theorem

The algorithm started with initial sentences $G \cup \{\neg \mathscr{S}\}$ will terminate with 'entailment is valid (invalid)' iff $G \models \mathscr{S}$ (not $G \models \mathscr{S}$).

4.4 Semantic tableaux for predicate calculus

We now turn to the more complex problem of applying semantic tableaux to predicate calculus.

Essentially we use exactly the same ideas that we used for propositional logic with two main extensions to deal with quantifiers.

Consider the sentence

$$\exists x(A(x) \vee B(x)) \rightarrow (\exists x A(x) \vee \exists x B(x))$$

which is valid as it stands, that is

$$\models \exists x(A(x) \vee B(x)) \rightarrow (\exists x A(x) \vee \exists x B(x))$$

We go about proving this by first negating it. The rule we need to apply first, since we have negated, and inside the negation $\rightarrow$ is the dominant connective, is the $\neg - \rightarrow$ rule.

After we have applied it the tableau is as in Figure 4.10. Now apply $\neg - \vee$ to (2) to get Figure 4.11.

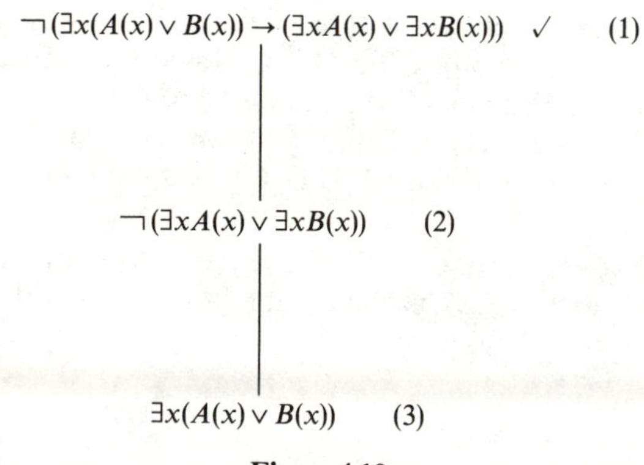

Figure 4.10.

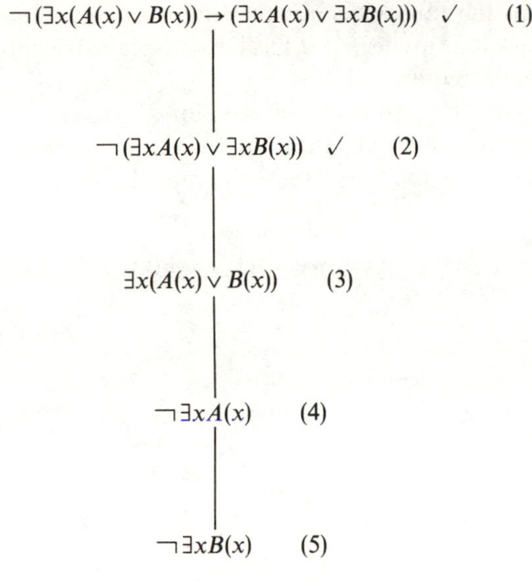

Figure 4.11.

The next stage is to move the negations of (4) and (5) inwards. We do this by using the equivalences

$$\neg\exists x A \leftrightarrow \forall x \neg A$$
$$\neg\forall x A \leftrightarrow \exists x \neg A$$

which means that we add the fragment

$$\forall x \neg A(x) \qquad (6)$$
$$\forall x \neg B(x) \qquad (7)$$

to the tableau.

Now we are in the position of having no more rules to apply since the connective in (3) is not dominant. This is where we need some rules for dealing with $\forall$ and $\exists$.

Remember that we are aiming to display a tableau in which every possible interpretation is given to any sentence appearing in it. Thus, when we meet a sentence saying 'there exists something such that …' it needs a name to form an instance. However, it says that this something denotes an object with specific attributes, that is, in this case, A is true of it or B is true of it or both. Thus, to preclude imposing these attributes on an object already mentioned, as will be the case in general, and of which they may not be true, we use an as yet unused name. Let us choose b. Thus we now add

$$(A(b) \vee B(b))$$

as sentence (8) to the tableau. Since the sentence only demanded one such name we can mark it as used since it has been displayed according to our goal and that is all that is required.

Now we have (6) and (7) whch say 'for all things ...'. We must take every name which has so far appeared on the path, that is all objects that we know, so far, appear in the universe of the model that is being constructed,

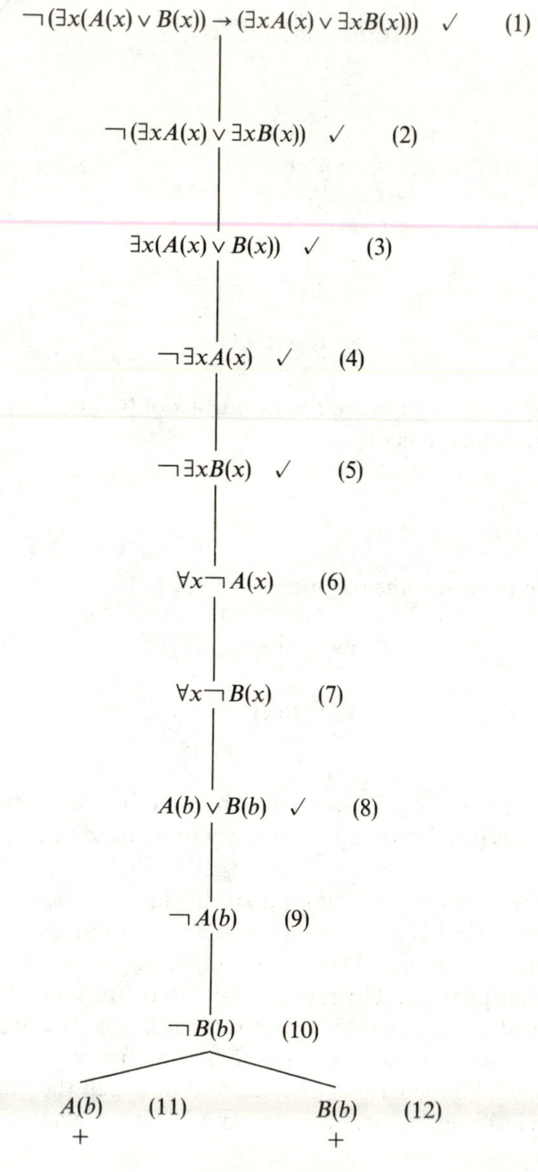

Figure 4.12.

which includes the sentence with which we are dealing, and place an instance of the sentence, for each name, on the path. This ensures that the given sentence has been interpreted in every way possible in the model as so far constructed. And this brings out an important point. At some future stage some new names may be added to the possible universe which we are gradually constructing.

Therefore, to keep to the goal of allowing every possible interpretation of every sentence we must allow the possibility that different instances of the sentence will be needed in the future. Thus we do not mark this sentence as having been used so that it may be taken into account in later stages. So we apply the ∀-rule to (6) and (7) with the name *b*. Then we can apply the ∨-rule to (8) to get the tableau so far as in Figure 4.12. We note that the paths both contain contradictions. So we can say that given the chance at each stage to produce a model we have failed. There is no model for our original negated sentence so

$$\exists x(A(x) \vee B(x)) \rightarrow (\exists x A(x) \vee \exists x B(x))$$

is valid.

4.5 Soundness and completeness for predicate calculus

We now collect together the rules we need and finally produce a simple algorithm for generating complete tableaux whenever possible in the predicate logic case.

Definition 4.11: The **universal rule** is:

> Given a sentence of the form $\forall x \mathscr{S}$ with x free in $\mathscr{S}$ on an open path which contains names $n_0, \ldots, n_{m-1} \in N$, add each of the sentences $\mathscr{S}[n_i/x]$ to the end of the path iff $\mathscr{S}[n_i/x]$ does not appear already on the path, where $0 \leqslant i \leqslant m-1$. Note for future use that the tableau has changed.

We note here that there may not be any names on the path when this rule is applied. In this case, since we are dealing with a non-empty universe, we take any name and apply the rule using this name. We shall usually take *a* to be this first name.

Definition 4.12: The **existential rule** is:

> Given a sentence of the form $\exists x \mathscr{S}$ with x free in $\mathscr{S}$ on an open path which contains names $n_0, \ldots, n_{m-1} \in N$, mark the sentence

as used and add the sentence $\mathscr{S}[n_m/x]$ where $n_m \in N \backslash \{n_0, \ldots, n_{m-1}\}$. Note for future use that the tableau has changed.

Consider the sentence

$$\exists y \forall x P(x, y) \tag{4.2}$$

This is not valid. Almost any interpretation you are likely to think up will show this. For example, let U = natural numbers, $\Psi(P)$ = less-than. Then the sentence says 'there is a largest natural number'.

Now consider the following. Since sentences of the form $\forall x \mathscr{S}$ are never marked we can always apply them to new names and a new name is always produced by an application of $\exists x \mathscr{S}$ so a sentence such as 4.2 will produce an infinite tableau which starts as in Figure 4.13 and will go on and on growing. In this case after any finite time the tableau is not closed and we still have some sentences to apply, and thus we continue. So, in our search for a countermodel it seems that we might, in some cases, never reach a conclusion since the process of building the tableau will never stop.

There are two points to make here. First, notice that by stepping outside of the algorithmic framework we were able to see that the sentence is in fact not valid. Thus, it is clear that algorithmic methods certainly are not the last word – which is comforting for humans (but not for expert systems salespeople). Secondly, this limitation is not specific to our method here.

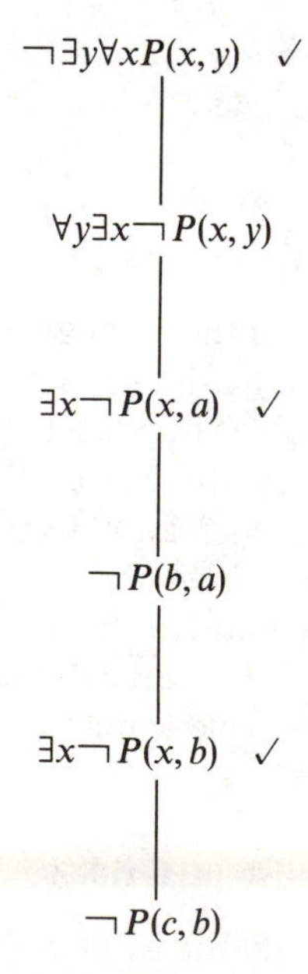

Figure 4.13.

Contrast this result with the propositional case where the tableau for any initial sentences always stops extending. This means that the question 'does G entail $\mathscr{S}$?' can always be answered either 'yes' or 'no'. We say that the validity question in propositional logic is decidable. It seems, however, that in the predicate calculus case we cannot hope for such a state of affairs.

But things are not as bad as they might seem. It is a central theorem of this subject that if a sentence is valid then there are methods for finding this out. In fact, the method of semantic tableaux is one such. We say that the validity question in predicate logic is **semidecidable**. The following informal argument should help to justify this.

We can extend the algorithm in the section above to get:

```
begin
  repeat
    changes:=false
    close each path that contains a sentence and its negation
    if all paths are closed
    then deliver "correct entailment"
    else
      if a connective rule can be applied
      then
        changes:=true
        apply the appropriate rule
        mark the sentence as used
      else
        apply the existential rule
        apply the universal rule
  until not changes
  deliver "incorrect entailment"
end
```

We can argue for the correctness of this algorithm in the same way as for the propositional case though with one big difference. Because a direct extension by the universal rule does not exhaust the sentence it is applied to, that is the sentence is not marked, we may never actually get a completed tableau in some cases. However, since any given universal sentence will always be dealt with after some finite time (since the preceding part of the algorithm deals only with the connectives, and they are always dealt with in a finite time) it is clear that any tableau makes progress towards being completed. So, even if some path is never closed, it must come nearer and nearer to forming a model set. (Smullyan (1968) and Jeffreys (1967) give other examples of how to organize the rules so that they always lead to a model set if one is possible.)

From this discussion we can show a soundness theorem and a completeness theorem as before in the propositional case, though we will not give them here since we are more interested in the computation side of the problem. However, see Smullyan (1968), for instance, for details.

> ***Correctness theorem***
>
> The algorithm started with initial sentences $G \cup \{\neg \mathscr{S}\}$ will terminate with 'entailment is valid' iff $G \models \mathscr{S}$. If the algorithm terminates with 'entailment is invalid' then not $G \models \mathscr{S}$. But, it may not terminate if not $G \models \mathscr{S}$.

What we are trying to do here is to establish theoremhood within a theory (which is a semantic notion, that is by choosing the 'right' substitution we are trying to decide on an assignment of elements of the universe to the variables which makes the sentences true by purely syntactic means. Of course, this has been done rather informally above, that is we have resorted rather a lot to woolly justification, but the aim of the chapter is to introduce a completely formal procedure which will decide theoremhood in a given theory.

Even though we have succeeded in this task, we should also point out that following the algorithm, while guaranteed to lead to the correct answer (if there is one) is not always the best strategy if you want a short refutation. As you will find if you try the exercises below, following the algorithm will lead to a far longer proof than will the strategy of looking at each stage of the development of the tableau and using your intelligence to plan how to proceed, keeping in mind that you want as short a refutation as possible. Another feature of this strategy, the first being that it requires intelligence and some idea of what a 'good' plan is, is that it gets better with practice, that is you also need to learn about what you are doing to be effective.

Now, our algorithm has none of these features and so we should expect it to be fairly ineffective for most of the time. However, since no one has yet succeeded in developing an algorithm which will learn in such a complex domain and which displays the sort of behaviour that we would call intelligent, we are left with largely ineffective algorithms. However, work in this area is developing new and (usually slightly) improved algorithms continuously. The most obvious strategy that we would like to add to our algorithm would give us some guidance as to which names to instantiate in universally quantified sentences in the universal rule. That is, instead of instantiating with every name, as we do here, is there a way of being sure of picking all and only those names needed still to give a sound and complete algorithm? The answer to this question is complicated and fairly subtle, so we will not go further into it here. Also, later in the book, in the logic programming chapter, we will address essentially the same problem in a different setting.

In Appendix B we give SML programs which extend the ones given earlier in this chapter and that implement the algorithm for the predicate case.

Exercise 4.2

(a) Use tableaux to decide on the status of the following entailments:

(i) $\{\forall y(H(y) \rightarrow A(y))\} \models$
$\forall x(\exists y(H(y) \wedge T(x, y)) \rightarrow \exists y(A(y) \wedge T(x, y)))$

(ii) $\models \forall x(P(x) \rightarrow \forall x P(x))$

(iii) $\models \exists x(P(x) \rightarrow \forall x P(x))$

(iv) $\models \exists x \forall y(S(x, y) \leftrightarrow \neg S(y, y))$

(b) How would you express (a(iv)) in English if S was the predicate 'x shaves y'?

(c) Express

'There is a man in town who shaves all the men in town who do not shave themselves' entails 'Some man in town shaves himself'

as an entailment using predicate calculus. Then, use the tableau method to decide its status.

SUMMARY

- *Semantic tableaux* provide another method of determining validity and consistency in predicate calculus. Tableaux are more concise and can be constructed more mechanically than proofs in axiom systems and, even for propositional logic, tableaux are much more efficient than truth-tables while giving exactly the same information.
- A tableau proof of $G \models \mathscr{S}$ starts with the set of sentences $G \cup \{\neg \mathscr{S}\}$. The tableau itself is a binary tree constructed from this initial set of sentences by using rules for each of the logical connectives and quantifiers that specify how the tree branches. If the tableau *closes* then the initial set of sentences is unsatisfiable and the entailment $G \models \mathscr{S}$ holds. A tableau closes if every branch closes. A branch closes if it contains $\mathscr{F}$ and $\neg \mathscr{F}$ for some sentence $\mathscr{F}$.
- If a tableau does not close, and yet is complete because none of the rules can be applied to extend it, then the initial sentences are satisfiable and the unclosed branches give a model for them. In this case $G \models \mathscr{S}$ does not hold. So a tableau proof is a proof by *refutation*; the existence of a model for $G \cup \{\neg \mathscr{S}\}$ is refuted.

- For the propositional calculus, semantic tableaux give a decision procedure (just as truth-tables do). For predicate calculus, the rule for universal sentences (sentences of the form $\forall x \mathscr{F}$) can be applied repeatedly. Methods of applying the rules can be given that are guaranteed to terminate with a closed tableau if and only if the initial set of sentences is unsatisfiable (the tableau method is *sound* and *complete*). However, if the initial set is satisfiable (the corresponding entailment is not valid) the method may never terminate. It can be shown that no method could terminate with the correct answer for all cases of invalidity. Predicate calculus is *semidecidable*.

5 Natural Deduction

5.1 Rules and proofs

Both of the methods given in previous chapters for constructing proofs have their disadvantages. Axiom systems are difficult to construct proofs in; their main uses are meta-logical, the small number of rules making it easier to prove results about logic.

The tableau method on the other hand is easy to use mechanically but, because of the form of the connective rules and the fact that a tableau starts from the negation of the formula to be proved, the proofs that result are not a natural sequence of easily justifiable steps. Likewise, very few proofs in mathematics are from axiom systems directly. In practice, mathematicians usually reason in a more flexible way.

5.1.1 Natural deduction rules

Suppose that, on the assumption that some statement P is true, Q can be shown to hold, possibly via some intervening proof steps. Since, given P, Q holds we can conclude (using the truth table for $\rightarrow$) that $P \rightarrow Q$ holds (given nothing). We can represent this sort of reasoning with a diagram

$$\frac{\begin{array}{c}\not{P}\\ \vdots \\ Q\end{array}}{P \rightarrow Q}$$

where P is crossed through to remind us that, once $P \rightarrow Q$ has been derived, P need no longer be considered an assumption; $P \rightarrow Q$ is true outright. We say that the original assumption P has been **discharged** in the process of going from Q to $P \rightarrow Q$.

What we have, in effect, is a method for introducing $\rightarrow$, generating a new formula with $\rightarrow$ as the principal connective. Similar introduction rules can be given for the other connectives and rules also for eliminating them. A set of rules that is adequate for deriving all theorems in predicate logic is given in Figure 5.1. The rules for $\wedge I$, $\wedge E$ and $\vee I$ are straightforward, while $\rightarrow E$ is modus ponens; however, $\vee E$ requires some explanation. The rule is essentially that of analysis by cases, familiar from mathematics. Suppose we want to show that some theorem is true for all integers. One method of proceeding, for example if the theorem has something to do with division by 2, would be to show that, if n is an even integer, the theorem holds, and then show that for n odd the theorem also holds. Since integers must be either odd or even the theorem therefore holds for all integers. The truth of the conclusion does not depend on whether n is odd or even (even though the form of the proof perhaps does) so we are justified in discharging the assumptions we made to get the proof.

In a similar way the rule C expresses another frequently used pattern of reasoning, commonly known as reductio ad absurdum or proof by contradiction. Given a result to be proved we assume that the result is not true and then show that falsity follows. Either the result is true or it is not; the assumption that it is not true leads to a contradiction so the result must be true. Note that the validity of this argument depends on accepting that the result is either true or not true and so one could argue that the rule is really a special case of $\vee E$ applied to $P \vee \neg P$. This makes the point that the set of rules we have given is not the only satisfactory set that could be given. We could have taken $P \vee \neg P$ as fundamental and got the effect of the rule C from this and $\vee E$. The important thing about a set of rules is that it should be sound and complete. Sound because, by using the rules, you cannot do other than derive valid conclusions, and complete because, if you are clever enough to see how, you can derive any valid conclusion. The set of rules that we give is both sound and complete, although we cannot give the rather technical proof of this here.

For aesthetic reasons we also like our rules to be minimal, in the sense that if we left out any rule the set would no longer be complete. We carry the tools for the job, but we carry no more than strictly necessary. Naturally this

$$\wedge I\frac{\mathscr{S} \qquad \mathscr{T}}{\mathscr{S} \wedge \mathscr{T}} \qquad \wedge E\frac{\mathscr{S} \wedge \mathscr{T}}{\mathscr{S}} \qquad \wedge E\frac{\mathscr{S} \wedge \mathscr{T}}{\mathscr{T}}$$

$$\vee I\frac{\mathscr{S}}{\mathscr{S} \vee \mathscr{T}} \qquad \vee I\frac{\mathscr{T}}{\mathscr{S} \vee \mathscr{T}} \qquad \vee E\frac{\mathscr{S} \vee \mathscr{T} \quad \begin{array}{c}\not{\mathscr{S}}\\ \vdots\\ \mathscr{R}\end{array} \quad \begin{array}{c}\not{\mathscr{T}}\\ \vdots\\ \mathscr{R}\end{array}}{\mathscr{R}}$$

$$\to I\frac{\begin{array}{c}\not{\mathscr{S}}\\ \vdots\\ \mathscr{T}\end{array}}{\mathscr{S} \to \mathscr{T}} \qquad \to E\frac{\mathscr{S} \to \mathscr{T} \qquad \mathscr{S}}{\mathscr{T}} \qquad \bot\frac{\bot}{\mathscr{S}} \qquad C\frac{\begin{array}{c}\neg\not{\mathscr{S}}\\ \vdots\\ \bot\end{array}}{\mathscr{S}}$$

$$\forall I\frac{\mathscr{S}(a)}{\forall x \mathscr{S}(x)}$$ provided that a does not occur in $\mathscr{S}(x)$ or any premise on which $\mathscr{S}(a)$ may depend

$$\forall E\frac{\forall x \mathscr{S}(x)}{\mathscr{S}(a)} \qquad \exists I\frac{\mathscr{S}(a)}{\exists x \mathscr{S}(x)}$$

$$\exists E\frac{\exists x \mathscr{S}(x) \quad \begin{array}{c}\not{\mathscr{S}(a)}\\ \vdots\\ \mathscr{T}\end{array}}{\mathscr{T}}$$ provided that a does not occur in $\mathscr{S}(x)$ or $\mathscr{T}$ or any assumption other than $\mathscr{S}(a)$ on which the derivation of $\mathscr{T}$ from $\mathscr{S}(a)$ depends

Notes:

(1) $\mathscr{S}$, $\mathscr{T}$ and $\mathscr{R}$ stand for any formula
(2) $\neg\mathscr{S}$ is defined as $\mathscr{S} \to \bot$

Figure 5.1.

places a burden on our ingenuity and as we gain experience of using the rules we add to our stock of tools in the form of lemmas which, just as in mathematics, are small but useful results that we often find ourselves using. However, further discussion of strategy and technique must come later. For the moment we continue our discussion of the rules.

It will be seen that we have chosen to define negation in terms of $\bot$, which it will be recalled is the symbol for a proposition that is false in every valuation. $P \to \bot$ is false when P is true and true when P is false, so $P \to \bot$ and $\neg P$ are logically equivalent. In the natural deduction system that we are using, $\bot$ is fundamental and $P \to \bot$ is the way you express negation, but for clarity and brevity we will allow negated formulas to appear in proofs on the understanding that $\neg P$ stands at all points for $P \to \bot$. There are implicit introduction and elimination rules for negation because we have as instances of $\to I$ and $\to E$ the rules

$$\frac{\begin{array}{c}\not{P}\\ \vdots\\ \bot\end{array}}{P \to \bot} \qquad \frac{P \quad P \to \bot}{\bot}$$

Our rule $\bot$ comes from the fact that an inconsistent set of formulas entails any formula. Note the difference between this and the rule C. The latter allows you to discharge the formula $\neg P$, the former does not.

Turning to the rules for the quantifiers, those for $\forall E$ and $\exists I$ are clearly sound, whatever the constant a denotes. The rule for $\forall I$ depends on the constant a being 'typical'; hence the condition.

Mathematical proofs often start 'take any integer a ...' and conclude 'but the choice of a was arbitrary and so the result holds for all a'. If it turns out that the result about a (that is $\mathscr{S}(a)$ in the rule) depends on a being prime, say, then you are not justified in concluding that the result holds for all a. The rule for $\exists E$ contains a similar caveat. Suppose that $\mathscr{S}(x)$ is some mathematical equation. We let the constant a be a solution to the equation and derive some other result $\mathscr{T}$. Now if there are other conditions that a has to satisfy and the derivation of $\mathscr{T}$ from $\mathscr{S}(a)$ depends on these as well, then it may turn out that no solution to the equation $\mathscr{S}(x)$ satisfies these further conditions and so the proof is not valid.

It will be apparent from what we have said that, unlike the axiom and tableau systems already discussed, these rules are designed to mimic fairly closely the patterns of reasoning that we use in both mathematics and everyday argument. For this reason the system is known as **natural deduction**. Its use in computer science has been found particularly appropriate where people collaborate with computers in an interactive way to construct proofs about programs. The LCF system (which we describe in Chapter 9), for example, is based on natural deduction. Natural though the rules may be, they still form part of a formal system in that they enable us to manipulate formulas in a purely syntactic way without regard to their meaning. As we indicated above the deductive system based on the rules can be shown sound and complete with respect to the semantics we gave in Chapter 3. They are therefore logically equivalent to the axiom and tableau methods covered already.

Natural deduction proofs are constructed by fitting the rules together to form a tree with the required formula at the root. Here for example is a demonstration of $\vdash P \to (Q \to (P \land Q))$.

$$\to I\,\frac{\to I\,\dfrac{\land I\,\dfrac{\not{P}^2 \qquad \not{Q}^1}{P \land Q}}{Q \to (P \land Q)}\,1}{P \to (Q \to (P \land Q))}\,2$$

The application of a rule is indicated by a horizontal line, on the left of which is the name of the rule used. For instance, in the example above, the first rule applied is $\wedge I$ (which in discussing it with someone else you would pronounce 'and introduction'). Assumptions are identified with numbers as they are introduced and when they are discharged, that is used in a rule such as $\rightarrow I$ which brings down the assumption as the antecedent of an introduced implication, then the corresponding number is written to the right of the discharging rule line and the assumption is crossed through to show that it has been discharged. Note that the rule names and numbers are not part of the proof; they are just annotations to make it easier to check that the rules have been followed.

Ideally you would have seen this proof develop as we wrote it down, while listening to the verbal explanation. Viewing the actual performance would also have given you some idea of how we went about inventing the proof, that is the proof strategy, as computer scientists call it. From the finished article you cannot see whether we started at the top, the bottom or the middle. So here is the action replay.

We know that we have to end up with $P \rightarrow (Q \rightarrow (P \wedge Q))$ as the bottom line, so what could the rule application that gives this be? For the moment we are working backwards. We start by writing down

$$?\,\frac{?}{P \rightarrow (Q \rightarrow (P \wedge Q))}$$

Now $\wedge E$ and $\vee E$ are not very likely because you would be coming from a more complicated formula that nevertheless still contains the formula you want. C is a possibility, but we know by experience that this is something to try when all else fails (as we see several times later). The obvious choice is the introduction rule for the principal connective, namely $\rightarrow I$. So we write down

$$\begin{array}{c} \cancel{P}^{1} \\ \\ ? \\ \\ \rightarrow I\,\dfrac{Q \rightarrow (P \wedge Q)}{P \rightarrow (Q \rightarrow (P \wedge Q))}\,1 \end{array}$$

Now we are in the same position again and the same argument gives

$$\begin{array}{c} \cancel{P}^{1} \qquad\qquad \cancel{Q}^{2} \\ \\ ? \\ \\ \rightarrow I\,\dfrac{\rightarrow I\,\dfrac{P \wedge Q}{Q \rightarrow (P \wedge Q)}\,2}{P \rightarrow (Q \rightarrow (P \wedge Q))}\,1 \end{array}$$

Now we see that things have worked out well for us because we have all the ingredients for an instance of $\wedge I$, so we complete the proof as shown above.

Of course it does not always work out so easily. It can often be hard to see how to put the rules together to get the result you want, and even harder to discharge legally assumptions that you find it convenient to make. Ingenuity is often required, but this must be confined to fitting the rules together and not to making up new rules of your own. For example we have seen in Chapter 3 that $\neg\forall x P(x) \vdash \exists x \neg P(x)$. A common mistake is to 'prove' this by natural deduction with

$$\exists I \frac{\dfrac{\neg \forall x P(x)}{\neg P(a)}}{\exists x \neg P(x)}$$

the point being that there is no rule among those we have given that sanctions the passage from $\neg\forall x P(x)$ to $\neg P(a)$. (We give the correct proof below.)

5.1.2 Fitch box presentation

Before we look at some more examples we will just say something about presentation. We have chosen to show proofs as trees (albeit growing the right way up in contrast to the usual convention in computer science) but there are other ways of presenting natural deduction proofs that are sometimes more convenient typographically. The proofs are the same, only the presentation differs. One of these variants is due to Fitch. The idea is reminiscent of Chinese boxes, boxes within boxes. When you make an assumption you put a box round it. When you discharge the assumption the box is closed and the derived formula is written outside the box. You can imagine the box, if you like, as a machine, literally a black box, which verifies the formula and 'outputs' it. The box is part of the world outside, so you can copy into the box any formula from outside it, but no formula from inside the box, which of course may depend on assumptions local to the box, may be copied to the outside. Figure 5.2 shows the example above done in the Fitch style.

5.1.3 Natural and non-natural reasoning

One of the best methods of formulating proofs by natural deduction is to imagine how you would convince someone else, who did not know any formal logic, of the validity of the entailment you are trying to demonstrate. Here is an example with quantifiers. We want to show

$$\{\forall x(F(x) \rightarrow G(x)), \forall x(G(x) \rightarrow H(x))\} \vdash \forall x(F(x) \rightarrow H(x))$$

P	Assumption
Q	Assumption
P	From outer box
$P \wedge Q$	$\wedge I$
$Q \rightarrow (P \wedge Q)$	$\rightarrow I$
$P \rightarrow (Q \rightarrow (P \wedge Q))$	$\rightarrow I$

Figure 5.2.

If you were trying to convince someone of the validity of this you might say

Take an arbitrary object a
Suppose a is an F
Since all Fs are Gs, a is a G
Since all Gs are Hs, a is an H
So if a is an F then a is an H
But this argument works for any a
So all Fs are Hs

Note how the natural deduction proof in Figure 5.3 expresses this argument almost exactly.

Not all natural deduction proofs are as natural as this. One of the disconcerting things at first about natural deduction is that some simple tautologies can be tricky to derive. This is nearly always because they involve the C rule in one form or another. Figure 5.4 shows a derivation of $\vdash P \vee \neg P$. (Remember that $\neg P$ is defined as $P \rightarrow \bot$.)

Once again it helps to consider how this proof works at the informal level. Remember you cannot just say 'either P is true or P is false and so $\neg P$ is true' because this is what you are trying to prove! We are not going to get anywhere by $\vee I$ because this would involve proving either P or $\neg P$ from no assumptions, so we start by assuming the result is false and use the C rule.

$$\forall I\,\dfrac{\rightarrow I\,\dfrac{\rightarrow E\,\dfrac{\rightarrow E\,\dfrac{\cancel{F(a)}^{1} \qquad \forall E\,\dfrac{\forall x(F(x) \rightarrow G(x))}{F(a) \rightarrow G(a)}}{G(a)} \qquad \forall E\,\dfrac{\forall x(G(x) \rightarrow H(x))}{G(a) \rightarrow H(a)}}{H(a)}\,1}{F(a) \rightarrow H(a)}}{\forall x(F(x) \rightarrow H(x))}$$

Figure 5.3.

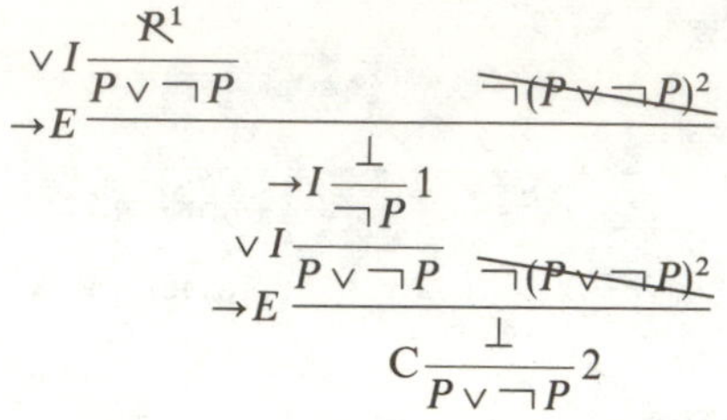

$$\cfrac{\cfrac{\cfrac{\cfrac{\cfrac{\cfrac{\cancel{P}^1}{P \vee \neg P}\,{\vee I} \qquad \cancel{\neg(P \vee \neg P)}^2}{\bot}\,{\to E}}{\neg P}\,{\to I}\ 1}{P \vee \neg P}\,{\vee I} \qquad \cancel{\neg(P \vee \neg P)}^2}{\bot}\,{\to E}}{P \vee \neg P}\,\mathrm{C}\ 2$$

Figure 5.4.

Now suppose P is true, then by the rule for $\vee I$ we have $P \vee \neg P$ and hence a contradiction. So $\neg P$ must be true (by definition of $\neg$, the assumption of P has led to a contradiction). But again by $\vee I$ we have $P \vee \neg P$ and hence once more a contradiction; however the difference this time is that P is no longer an outstanding assumption.

At this point the only current assumption is $\neg(P \vee \neg P)$, so discharging this with the C rule we get the required result. In fact we obtained $P \vee \neg P$ in three places but only on the last occasion did it not depend on an undischarged assumption. One final point of presentation: we introduced $\neg(P \vee \neg P)$ as an assumption on two separate occasions but gave it the same number each time and discharged both occurrences at once. This is quite sound. It is only a single assumption written twice for purposes of presentation. It would be a different matter though if we had discharged the first occurrence before introducing the second.

We could of course have included $P \vee \neg P$ as a primitive rule and you may sometimes see natural deduction systems in which this or something similar is done. However, now that we have carried out the proof above, it is clear that any formula could have been consistently substituted for P and both the proof and the result would be equally valid. So in, for example, the exercises that follow you may assume that this has been done and quote an appropriate instance of $P \vee \neg P$ as a lemma.

Finally, observe that at the last step we could have used $\to I$ to get $\neg\neg(P \vee \neg P)$, showing that this can be derived without $\bot$ or C. This is an important point that we return to when we look at other logics.

Now here are some examples with quantifiers. First, in Figure 5.5, we show how to derive the basic equivalences between $\forall$ and $\exists$, starting with $\{\neg \exists x P(x)\} \vdash \forall x \neg P(x)$.

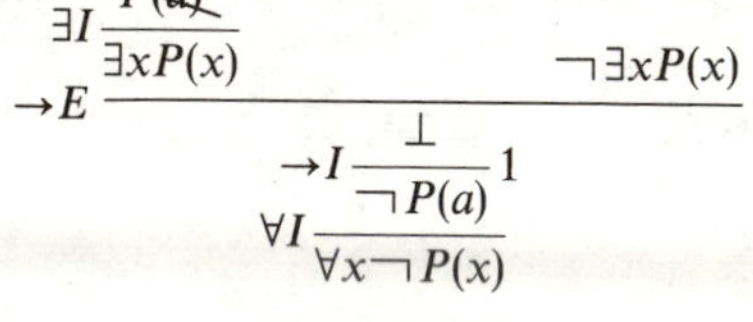

$$\cfrac{\cfrac{\cfrac{\cfrac{\cancel{P(a)}^1}{\exists x P(x)}\,{\exists I} \qquad \neg \exists x P(x)}{\bot}\,{\to E}}{\neg P(a)}\,{\to I}\ 1}{\forall x \neg P(x)}\,{\forall I}$$

Figure 5.5.

$$
\dfrac{\dfrac{\dfrac{\dfrac{\dfrac{\dfrac{[\neg P(a)]^1}{\exists x\neg P(x)}\,\exists I \qquad [\neg\exists x\neg P(x)]^2}{\bot}\,\to E}{P(a)}\,C\ 1}{\forall x P(x)}\,\forall I \qquad \neg\forall x P(x)}{\bot}\,\to E}{\exists x\neg P(x)}\,C\ 2
$$

Figure 5.6.

Note that we have labelled the line marking the second rule application as $\to E$ because the $\neg\exists xP(x)$ in the line above is a synonym for $\exists xP(x) \to \bot$, and a similar convention is used in reverse in the following line.

Two of the remaining three facts of this type, namely $\{\forall x\neg P(x)\} \vdash \neg\exists xP(x)$ and $\{\exists x\neg P(x)\} \vdash \neg\forall xP(x)$, are straightforward and we leave them as exercises. However, $\{\neg\forall xP(x)\} \vdash \exists x\neg P(x)$ requires use of the C rule and its derivation is given in Figure 5.6.

Finally Figure 5.7 shows an example of a fallacious 'proof' of $\{\exists xF(x), \exists xG(x)\} \vdash \exists x(F(x) \wedge G(x))$ to illustrate the need for the restriction on $\exists E$.

$$
\dfrac{\exists x G(x) \qquad \dfrac{\exists x F(x) \qquad \dfrac{\dfrac{[F(a)]^1 \qquad [G(a)]^2}{F(a)\wedge G(a)}\,\wedge I}{\exists x(F(x)\wedge G(x))}\,\exists I}{\exists x(F(x)\wedge G(x))}\,\exists E\ 1}{\exists x(F(x)\wedge G(x))}\,\exists E\ 2
$$

Figure 5.7.

The error is in the topmost use of $\exists E$. At that point the constant a occurs in an undischarged assumption, $G(a)$, on which the proof of $\exists x(F(x) \wedge G(x))$ at that stage still depends. A similar example can be given to show the need for the condition in $\forall I$, and we leave it as an exercise for the reader to construct one.

Exercise 5.1 Show using the natural deduction rules given that the following are correct:

(a) $\vdash (P \to (Q \to R)) \to ((P \to Q) \to (P \to R))$

(b) $\vdash \neg\neg P \to P$

(c) $\vdash (\neg P \to Q) \to (\neg Q \to P)$

(d) $\vdash (\forall xP(x) \to Q) \to \exists x(P(x) \to Q)$

(e) $\{\neg P\} \vdash P \to Q$

(f) $\{P \vee Q, \neg P\} \vdash Q$

(g) $\{\forall x \neg P(x)\} \vdash \neg \exists x P(x)$

(h) $\{\exists x \neg P(x)\} \vdash \neg \forall x P(x)$

(i) $\{P \rightarrow Q\} \vdash \neg Q \rightarrow \neg P$

5.2 The sequent calculus

5.2.1 Natural deduction rules as sequents

The tree format in which we have displayed natural deduction proofs is good for constructing proofs but not so good for reasoning about them. Suppose we want to demonstrate, for example, that from a derivation $\{A, B\} \vdash C$ we can get $\{A\} \vdash B \rightarrow C$ where A, B and C are any formulas; then we have to draw pictures to point out that a proof tree such as

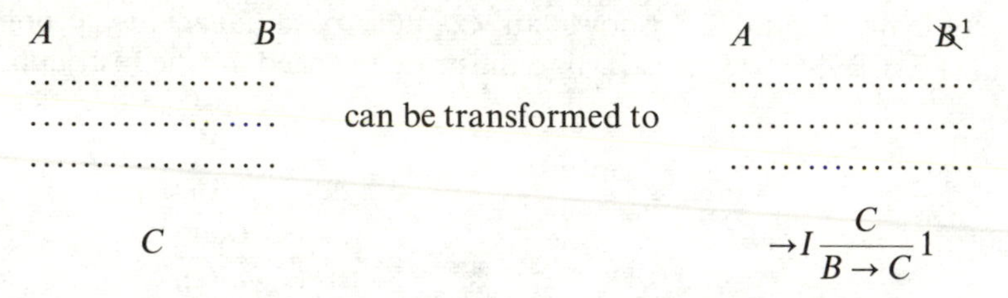

What we need is a less pictorial, more formal, notation for manipulating derivations. This is provided by the notion of a **sequent**. A sequent is an expression of the form $G \Rightarrow H$ where H and G are sets of formulas. To start with we shall be talking about cases in which H is a single formula, A say, in which case we shall allow ourselves to write $G \Rightarrow \{A\}$ as $G \Rightarrow A$. We shall also for brevity write $G \cup \{A\} \Rightarrow B$ as $G, A \Rightarrow B$ and adopt the convention that $G \Rightarrow \{\ \}$, $G \Rightarrow$ and $G \Rightarrow \bot$ are all the same.

The use of the $\Rightarrow$, instead of the $\vdash$ that you might have expected, is designed to make an important distinction clear. If you assert $G \vdash B$ you are saying, in the observer's language, that B can be derived from the set of assumptions G in some system of deduction, whereas $G \Rightarrow B$ is just a formula in the observer's language. Like a formula such as $x + 7$ in ordinary algebra, it does not make sense to say that it holds or does not hold – it is just there to be manipulated. We are going to put forward a set of rules that say how formulas involving $\Rightarrow$ can be manipulated and we will draw conclusions of the form 'if $G \Rightarrow A$ can be obtained by using the rules then $G \vdash A$ in classical logic'.

You may notice a similarity between this and the kind of statement we were making in the axiom systems we looked at earlier. This is no coincidence. In Chapter 2 we were defining the notion of proof in a formal system; here we

$$\text{Assumption}\ \frac{}{\mathscr{S} \Rightarrow \mathscr{S}}$$

$$\text{Thin}\ \frac{H \Rightarrow \mathscr{S}}{G, H \Rightarrow \mathscr{S}}$$

$$\wedge I\ \frac{G \Rightarrow \mathscr{S} \quad H \Rightarrow \mathscr{T}}{G, H \Rightarrow \mathscr{S} \wedge \mathscr{T}} \qquad \wedge E\ \frac{G \Rightarrow \mathscr{S} \wedge \mathscr{T}}{G \Rightarrow \mathscr{S}} \qquad \wedge E\ \frac{G \Rightarrow \mathscr{S} \wedge \mathscr{T}}{G \Rightarrow \mathscr{T}}$$

$$\vee I\ \frac{G \Rightarrow \mathscr{S}}{G \Rightarrow \mathscr{S} \vee \mathscr{T}} \qquad \vee I\ \frac{G \Rightarrow \mathscr{T}}{G \Rightarrow \mathscr{S} \vee \mathscr{T}} \qquad \bot\ \frac{G \Rightarrow \bot}{G \Rightarrow \mathscr{S}}$$

$$\vee E\ \frac{G \Rightarrow \mathscr{S} \vee \mathscr{T} \quad \mathscr{S}, H \Rightarrow \mathscr{R} \quad \mathscr{T}, F \Rightarrow \mathscr{R}}{G, H, F \Rightarrow \mathscr{R}} \qquad C\ \frac{\neg \mathscr{S}, G \Rightarrow \bot}{G \Rightarrow \mathscr{S}}$$

$$\rightarrow I\ \frac{\mathscr{S}, G \Rightarrow \mathscr{T}}{G \Rightarrow \mathscr{S} \rightarrow \mathscr{T}} \qquad \rightarrow E\ \frac{G \Rightarrow \mathscr{S} \rightarrow \mathscr{T} \quad H \Rightarrow \mathscr{S}}{G, H \Rightarrow \mathscr{T}}$$

$$\forall I\ \frac{G \Rightarrow \mathscr{S}(a)}{G \Rightarrow \forall x \mathscr{S}(x)}$$ provided that a does not occur in G or $\mathscr{S}(x)$

$$\forall E\ \frac{G \Rightarrow \forall x \mathscr{S}(x)}{G \Rightarrow \mathscr{S}(a)} \qquad \exists I\ \frac{G \Rightarrow \mathscr{S}(a)}{G \Rightarrow \exists x \mathscr{S}(x)}$$

$$\exists E\ \frac{G \Rightarrow \exists x \mathscr{S}(x) \quad \mathscr{S}(a), H \Rightarrow \mathscr{T}}{G, H \Rightarrow \mathscr{T}}$$ provided that a does not occur in H, $\mathscr{S}(x)$ or $\mathscr{T}$

Figure 5.8.

are formalizing the notion of entailment. Note by the way that $\Rightarrow$ does not necessarily have anything to do with the symbol $\rightarrow$ in the language of predicate calculus, and there will be nothing in our rules that results in formulas such as $A \Rightarrow (B \Rightarrow C)$ being obtained. The symbol $\Rightarrow$ will never be iterated.

Figure 5.8 shows the natural deduction rules expressed in sequent notation. Note how the sets of formulas on the left of the $\Rightarrow$ are used to bundle together and to make explicit the assumptions that were part of our verbal explanation above (F, G, H are sets of formulas).

To make these complete we have had to add (at the beginning) two more rules that were also previously part of our verbal explanation. The first of them (which has been given the name 'Assumption') gives you a way of getting started. You might think of it as (the only) axiom schema of this formal system for writing down derivations. The second is called 'Thin' and the remainder are direct translations of the natural deduction rules we had earlier.

As an example the very first natural deduction derivation of the chapter can be rewritten using the sequent form of the rules as in Figure 5.9.

$$\to I\ \dfrac{\to I\ \dfrac{\wedge I\ \dfrac{\text{Assumption}\ \dfrac{}{P \Rightarrow P} \qquad \text{Assumption}\ \dfrac{}{Q \Rightarrow Q}}{P, Q \Rightarrow P \wedge Q}}{P \Rightarrow Q \to (P \wedge Q)}}{\Rightarrow P \to (Q \to (P \wedge Q))}$$

Figure 5.9.

We implied above that one of the motives behind rewriting natural deduction in this form is to prove meta-theorems about natural deduction. Here is an example. If G and H are sets of formulas and $\mathcal{S}$ and $\mathcal{T}$ are any formulas we fit the rules together to give

$$\to E\ \dfrac{G \Rightarrow \mathcal{S} \qquad \to I\ \dfrac{\mathcal{S}, H \Rightarrow \mathcal{T}}{H \Rightarrow \mathcal{S} \to \mathcal{T}}}{G, H \Rightarrow \mathcal{T}}$$

We have shown that, whatever G, H, $\mathcal{S}$ and $\mathcal{T}$ are, from $G \Rightarrow \mathcal{S}$ and $\mathcal{S}, H \Rightarrow \mathcal{T}$ we can get $G, H \Rightarrow \mathcal{T}$. This derived rule is called the **cut rule** and, for $G = H$, will be recalled from Chapter 2 where we said that if $G \models \mathcal{S}$ and $G, \mathcal{S} \models \mathcal{T}$ then $G \models \mathcal{T}$, where $\models$ denotes semantic entailment.

Exercise 5.2 Show that

$$\dfrac{G \Rightarrow \neg\neg\mathcal{S}}{G \Rightarrow \mathcal{S}}$$

is also a derived rule of this system.

5.2.2 Classical sequent calculus

Sequents were introduced by Gentzen (1934) who proposed a calculus in which elimination rules are replaced by introduction rules on the left of the $\Rightarrow$. Gentzen proved the equivalence of his sequent calculus to natural deduction by adding the cut rule to his system, which makes relating it to natural deduction a lot easier. Then, in a famous result known as Gentzen's *Hauptsatz*, he showed that any derivation involving the cut rule could be converted to one that was cut free.

Gentzen's sequent calculus idea is a very powerful one for studying derivations and relating them to other systems such as the tableau method. A further generalization allows more than one formula on the right of the $\Rightarrow$. Doing this, and bringing in rules for $\neg$, gives the symmetrical system shown in Figure 5.10 which again can be shown equivalent to the natural deduction system for full predicate logic.

$$\text{Assumption}\ \frac{}{\mathscr{S} \Rightarrow \mathscr{S}}$$

$$\Rightarrow\text{Thin}\ \frac{G \Rightarrow H}{G \Rightarrow H, \mathscr{T}} \qquad \text{Thin}\Rightarrow\ \frac{G \Rightarrow H}{\mathscr{S}, G \Rightarrow H}$$

$$\Rightarrow\wedge\ \frac{G \Rightarrow H, \mathscr{S} \qquad G' \Rightarrow H', \mathscr{T}}{G', G \Rightarrow \mathscr{S} \wedge \mathscr{T}, H, H'} \qquad \wedge\Rightarrow\ \frac{\mathscr{S}, \mathscr{T}, G \Rightarrow H}{\mathscr{S} \wedge \mathscr{T}, G \Rightarrow H}$$

$$\Rightarrow\vee\ \frac{G \Rightarrow H, \mathscr{S}}{G \Rightarrow H, \mathscr{S} \vee \mathscr{T}} \qquad \Rightarrow\vee\ \frac{G \Rightarrow H, \mathscr{T}}{G \Rightarrow H, \mathscr{S} \vee \mathscr{T}} \qquad \vee\Rightarrow\ \frac{\mathscr{S}, G \Rightarrow H \qquad \mathscr{T}, G' \Rightarrow H'}{\mathscr{S} \vee \mathscr{T}, G, G' \Rightarrow H, H'}$$

$$\Rightarrow\rightarrow\ \frac{\mathscr{S}, G \Rightarrow H, \mathscr{T}}{G \Rightarrow H, \mathscr{S} \rightarrow \mathscr{T}} \qquad \rightarrow\Rightarrow\ \frac{G \Rightarrow H, \mathscr{S} \qquad \mathscr{T}, G' \Rightarrow H'}{\mathscr{S} \rightarrow \mathscr{T}, G, G' \Rightarrow H, H'}$$

$$\Rightarrow\neg\ \frac{\mathscr{S}, G \Rightarrow H}{G \Rightarrow H, \neg\mathscr{S}} \qquad \neg\Rightarrow\ \frac{G \Rightarrow H, \mathscr{S}}{\neg\mathscr{S}, G \Rightarrow H}$$

$$\Rightarrow\forall\ \frac{G \Rightarrow H, \mathscr{S}(a)}{G \Rightarrow H, \forall x \mathscr{S}(x)} \qquad \forall\Rightarrow\ \frac{\mathscr{S}(a), G \Rightarrow H}{\forall x \mathscr{S}(x), G \Rightarrow H}$$

$$\Rightarrow\exists\ \frac{G \Rightarrow H, \mathscr{S}(a)}{G \Rightarrow H, \exists x \mathscr{S}(x)} \qquad \exists\Rightarrow\ \frac{\mathscr{S}(a), G \Rightarrow H}{\exists x \mathscr{S}(x), G \Rightarrow H}$$

Note: in the rules for $\Rightarrow\forall$ and $\exists\Rightarrow$ a must not be free in G, H or $\mathscr{S}(x)$

Figure 5.10.

5.2.3 Tableaux and sequents

It can be shown that the statement '$G \Rightarrow H$ can be derived in the system above' is equivalent to the statement 'if all the formulas in G are true in an interpretation then at least one of the formulas in H is true'. This can be used as the basis for a demonstration that the tableau and sequent calculus are equivalent. In fact closed tableaux are essentially sequent derivations in this system written upside down.

For instance, in Figure 5.11 we have the tableau proof of $\{A \rightarrow B, B \rightarrow C\} \models A \rightarrow C$ and in Figure 5.12 a sequent proof of $\{A \rightarrow B, B \rightarrow C\} \Rightarrow A \rightarrow C$.

Now, if we read each sequent as describing the unused sentences on each path of the tableau, with sentences on the left of the $\Rightarrow$ being unnegated and those on its right being negated and with branches in the tableau matching branches in the sequent proof, the similarity becomes quite clear.

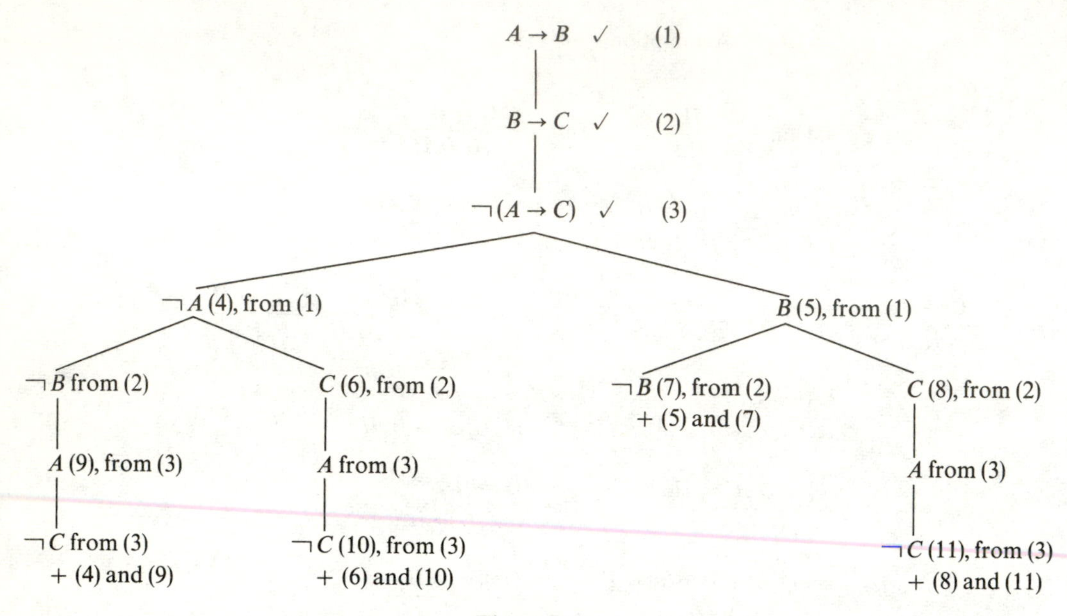

Figure 5.11.

So, starting at the bottom of the sequent proof (Figure 5.12) we have sentences (1), (2) and (3) from the tableau. Then, moving up the leftmost branch, we have sentences (2), (3) and (4), that is the unused ones on the corresponding tableau branch. Then, moving up leftmost again, we have sentences (3), (4) and the $\neg B$ from (2). Then we get, using $\Rightarrow\rightarrow$, sentences (4), the $\neg B$, (9) and the $\neg C$ from (3). This leads to a closure in the table of the leftmost path and, after some thinning, an assumption in the sequent proof.

The other paths can be seen to be equivalent to the corresponding tableau paths in the same way.

In fact, the original tableaux as introduced by Beth were even closer to the sequents since he used some notation to show whether, in sequent terms, a sentence was on the left or the right of the $\Rightarrow$. Our tableaux are a subsequent specialization of Beth's to make the presentation, at least for the classical logics, clearer.

$$
\dfrac{\dfrac{\dfrac{\dfrac{\dfrac{\dfrac{}{A \Rightarrow A}\text{Ass}}{A \Rightarrow A, B}\Rightarrow\text{Thin}}{A \Rightarrow A, B, C}\Rightarrow\text{Thin}}{\Rightarrow A, B, A \rightarrow C}\Rightarrow\rightarrow \quad \dfrac{\dfrac{\dfrac{\dfrac{}{A \Rightarrow A}\text{Ass}}{C, A \Rightarrow A}\text{Thin}\Rightarrow}{A, C \Rightarrow A, C}\Rightarrow\text{Thin}}{C \Rightarrow A, A \rightarrow C}\Rightarrow\rightarrow}{B \rightarrow C \Rightarrow A \rightarrow C, A}\rightarrow\Rightarrow \quad \dfrac{\dfrac{\dfrac{}{B \Rightarrow B}\text{Ass}}{B \Rightarrow B, A \rightarrow C}\Rightarrow\text{Thin} \quad \dfrac{\dfrac{\dfrac{\dfrac{}{C \Rightarrow C}\text{Ass}}{B, C \Rightarrow C}\text{Thin}\Rightarrow}{A, C, B \Rightarrow C}\text{Thin}\Rightarrow}{C, B \Rightarrow A \rightarrow C}\Rightarrow\rightarrow}{B \rightarrow C, B \Rightarrow A \rightarrow C}\rightarrow\Rightarrow}{A \rightarrow B, B \rightarrow C \Rightarrow A \rightarrow C}\rightarrow\Rightarrow
$$

Figure 5.12.

5.3 Generalizing the logic

5.3.1 Changing the rules to get a different logic

Up to this point we have only talked about a single logic, called classical logic, the logic of two truth values and a truth-functional valuation system as described in Chapter 2 and Chapter 3. In the introduction we promised to look not just at one logic but at many. The sequent calculus with its many rules, each intended to capture in a straightforward way the behaviour of the logical connectives, gives us the opportunity to do this.

We have already seen that without the rule called C we are apparently unable to give a natural deduction proof of the propositional tautology $P \vee \neg P$. We now show that this is not just due to lack of inventiveness on our part. It provides a good illustration of the use of the sequent calculus for reasoning about derivations. However, to avoid some detail, we ask the reader to take on trust that the natural deduction system without C is equivalent to the sequent calculus system in which for $\Rightarrow\rightarrow$, $\Rightarrow\neg$ and $\forall$ the set of formulas H on the right of the $\Rightarrow$ must be null.

We can derive $\neg\neg(P \vee \neg P)$ as shown in Figure 5.13. However, for any derivation of $\Rightarrow(P \vee \neg P)$ in this system, there would only be two possibilities at the last step. One is $\Rightarrow$ Thin, which is clearly not applicable, the other is $\Rightarrow\vee$, which would have either $\Rightarrow P$ or $\Rightarrow\neg P$ as an antecedent, so there is no way to derive $\Rightarrow(P \vee \neg P)$. As an exercise the reader should show that $\Rightarrow(P \vee \neg P)$ can be derived in the system of Figure 5.10 where H can be non-empty in the $\Rightarrow\neg$ rule.

5.3.2 Giving a semantics for the system without C

It is clear that, in any reasonable truth-functional valuation system based on two truth values, $P \vee \neg P$ has to take a single truth value in all valuations. If we are to give a semantics for a logical system in which $P \vee \neg P$ is not a theorem then the obvious solution is to admit more than two truth-values.

$$
\cfrac{\cfrac{\cfrac{\cfrac{\cfrac{\cfrac{P \Rightarrow P}{P \Rightarrow P \vee \neg P}\;\Rightarrow\vee}{P, \neg(P \vee \neg P) \Rightarrow}\;\neg\Rightarrow}{\neg(P \vee \neg P) \Rightarrow \neg P}\;\Rightarrow\neg}{\neg(P \vee \neg P) \Rightarrow P \vee \neg P}\;\Rightarrow\vee}{\neg(P \vee \neg P) \Rightarrow}\;\neg\Rightarrow}{\Rightarrow \neg\neg(P \vee \neg P)}\;\Rightarrow\neg
$$

Figure 5.13.

In fact many logics with multiple truth values have been invented to model various notions of uncertainty or to overcome apparent paradoxes in classical logic. To give some idea of the kind of reasoning that arises, and to provide some background for developments in Chapter 8, let us, in a similar way to that in Chapter 2, attempt to define, for the propositional case only, a more general valuation function $v_k\colon L \to \{1,\dots,k\}$ where $v_k(\top) = 1$ and $v_k(\bot) = k$ (Scott, 1981).

$$v_k(A \wedge B) = \max(v_k(A), v_k(B))$$
$$v_k(A \vee B) = \min(v_k(A), v_k(B))$$
$$v_k(A \rightarrow B) = 1 \text{ if } v_k(A) \geqslant v_k(B) \text{ and } k \text{ if } v_k(A) < v_k(B)$$

We then have

$$v_k(\neg A) = v_k(A \rightarrow \bot) = 1 \text{ if } v_k(A) = k \text{ and } k \text{ if } v_k(A) < k$$

It will be seen that v_2 reduces to the valuation function for classical propositional logic that we used in Chapter 2. We might hope that v_3 would provide an adequate valuation function for the system that you get by omitting the C rule.

Now that we are talking about more than one logical system we must be careful to distinguish between them. We will call the full set of natural deduction rules C (standing for either 'classical' or 'using contradiction'), and the system derived from the full set by deleting the C rule we shall call I, for a reason to be revealed shortly. If a formula A can be derived from a set of formulas G in the full natural deduction system C then we will write $G \vdash_C A$, and if A can be derived from G in the system I then we will write $G \vdash_I A$.

By a lengthy but straightforward induction over the natural deduction proof tree (there are seven cases, one for each of the propositional rules) we can show that if $G \vdash_I A$ then $\max\{v_k(t_i) \mid t_i \in T\} \geqslant v_k(A)$ and hence if $\vdash_I A$ then $v_k(A) = 1$. Now we can easily check that $v_2(P \vee \neg P) = 1$ for both possible values of $v_2(P)$, but $\vdash_I P \vee \neg P$ does not hold so v_2 is not adequate for I.

Constructing the 'truth-table' based on v_3 we get

$v_3(P)$	$v_3(\neg P)$	$v_3(P \vee \neg P)$
1	3	1
2	3	2
3	1	1

so the idea seems to have worked in this case. $P \vee \neg P$ is not a tautology in the three-valued semantics and so the fact that it is not a theorem of I does not disprove the adequacy of v_3.

Success is short lived, however, because it is possible to show that whatever value of k is chosen, there are formulas A for which $v_k(A) = 1$ in all valuations but it is not the case that $\vdash_I A$. In fact $(\neg\neg P \rightarrow P) \rightarrow (P \vee \neg P)$ is such a formula. We can leave it to the reader to show that $v_k(A) = 1$ for all k and all values of $v_k(P)$. To show that it is not the case that $\vdash_I A$ we have to wait until Chapter 8. The conclusion is that a simple extension of the truth-table idea with a linearly ordered set, however large, is insufficient to give a semantic basis for the system of deduction I. In Chapter 8 we show what the correct semantic basis is.

5.3.3 Constructive ideas

In fact I is a system of deduction for a logic known as intuitionistic logic. Historically it is important because it grew out of the work of Brouwer who originated the study of intuitionistic, for which you may read constructive, mathematics in the early part of this century. Brouwer rejected the law of the excluded middle, that is $\mathscr{S} \vee \neg\mathscr{S}$ for any sentence $\mathscr{S}$, and he and his followers set out to reconstitute mathematics on purely constructive lines. Constructivists would not, for example, admit existence proofs that rely on deriving a contradiction from the assumption that some mathematical entity does not exist. They said that for a proof of existence to have any meaning you have to actually produce an instance of the object in question. We can see an immediate connection here with the natural deduction system I.

In one of our earlier examples we showed that $\{\neg\forall x P(x)\} \vdash_C \exists x \neg P(x)$, which 'says' that, if it is not the case that every x has property P, then there must be some x that does not have property P. However, we did not do this by producing a particular object a such that $\neg P(a)$ was true. We assumed $\neg\exists x \neg P(x)$, derived $\bot$, and used the C rule. It is not hard to show that in the system I which does not have the C rule, you cannot prove $\exists x \neg P(x)$ from $\neg\forall x P(x)$.

It was Heyting around 1930 who first gave an informal semantics for intuitionistic logic in terms of the notion of constructive proof and Gentzen a few years later who essentially formulated the natural deduction systems for both I and C that we have been discussing in this chapter. We have just seen that a simplistic approach to giving the logic a mathematical semantics, using more than two truth-values, does not work. Tarski and Stone, working at about the same time as Gentzen, were the first to show the correct way to do it when they noticed similarities between intuitionistic logic and topology. However, Kripke (1963) proposed a semantics that is much easier to work with, and since we explain this in Chapter 8, we shall continue the discussion of intuitionistic logic there.

5.4 What is logic ultimately?

We made a point in our introduction of saying that we were going to look at not just one logic but many, and we have just shown you an example of this by examining the consequences of modifying the natural deduction rules for classical logic to get another important system. There is nothing sacred about classical logic. We can and will go on to look at other logics. But if we are going to do this where do we draw the line? Ultimately what is and what is not a logic?

Imagine I am showing you a computer program I have written. I place you in front of a terminal and invite you to type in things like

$$\{P, P \rightarrow Q\} \vdash Q$$

to which the machine answers 'yes', or

$$\{\neg P \rightarrow Q\} \vdash \neg Q \rightarrow P$$

to which the machine answers 'no', meaning that in whatever system of deduction my program implements, Q can be derived from P and $P \rightarrow Q$, but $\neg Q \rightarrow P$ cannot be derived from $\neg P \rightarrow Q$.

The symbol $\vdash$ denotes, in the observer's language, a relation, called a **consequence relation**, between formulas and sets of formulas of the object language. If you were to try out my program for a long time you would gradually build up some knowledge of what the consequence relation of my system was and you would start to get some idea of its properties. For example you might notice that it always said 'yes' to questions of the form $\{A\} \vdash A$, whatever formula A was, or that whenever it said 'yes' to $G \vdash A$ it would always say 'yes' to $G, B \vdash A$. Thinking back to Chapter 2 you recall that the relation of semantic entailment $\models$ possessed similar properties. If you observe that the program's consequence relation shares these properties then this increases your confidence that my program is implementing some kind of logic. This example is due to Dov Gabbay (1984).

Suppose, however, that I now tell you my program works by converting all the formulas to numbers and then says that $G \vdash A$ if the number corresponding to A, call it $n(A)$, divides the product $\prod n(g_i)$ of the numbers corresponding to the g_i in G. You would probably want to say that a program that worked on a purely arithmetic basis such as this did not have anything to do with logic. Yet if the program did work like this it would have at least some of the properties of the systems described above because $n(A)$ divides $n(A)$ and so $\{A\} \vdash A$ and if $n(A)$ divides $\prod n(g_i)$ then $n(A)$ divides $n(B)\prod n(g_i)$, and so if $G \vdash A$ then $G, B \vdash A$. If arithmetically based programs like this can have logical properties then where do we draw the line on deduction rules? What is and what is not logic?

A relation $\vdash$ between sets of formulas and formulas of some language L is the consequence relation of a logical deduction system based on L if it satisfies the three rules (where A and B are formulas and G and H are finite sets of formulas)

reflexivity	$\{A\} \vdash A$
monotonicity	$G \vdash A$ implies $G, B \vdash A$
transitivity (cut)	$G \vdash A$ and $A, H \vdash B$ implies $G, H \vdash B$

We leave the reader as an exercise to show by means of a counterexample that the arithmetically based relation above does not always satisfy the transitivity condition.

Nowadays, however, the definition of what constitutes logical deduction is being pushed wider and wider, partly as a result of the impetus given by computer science. For example, some logics used in artificial intelligence and the theory of computation do not satisfy the monotonicity condition.

SUMMARY

- Natural deduction is a formal deductive system that models ordinary mathematical reasoning more closely than axiom systems or the tableau method. A natural deduction system consists of rules for *introducing* and *eliminating* each of the logical connectives and quantifiers. Proofs are constructed by fitting the rules together in the form of a tree. As in ordinary reasoning, temporary assumptions may be made in the course of the proof and then *discharged* by incorporating them into the conclusion. Interactive proof systems for reasoning about programs in computer science are often based on natural deduction.
- The sequent calculus is a less pictorial, more algebraic, formulation of natural deduction in which the role of assumptions is more explicit. It provides a means of reasoning about proofs and axiomatizing deduction. A *sequent* is an expression of the form $G \Rightarrow H$ where G and H are sets of formulas. A sequent calculus is a set of rules for manipulating sequents. Gentzen gave a set of sequent rules for classical predicate calculus that he showed to be equivalent to natural deduction. The tableau method is shown to be essentially another way of writing sequent calculus derivations.
- Natural deduction and the sequent calculus give us the opportunity to generate different logics by varying the rules. Leaving out the contradiction rule we get a logic, called *intuitionistic logic*, in which the 'law of the excluded middle',

$P \vee \neg P$, is not a theorem. It is shown that this logic cannot have a semantics with two truth-values, or indeed any linearly ordered set of truth-values.

- If we can change the logic by changing the rules, how far can we go while still 'doing logic'? The rules of deduction determine a *consequence relation*. Consequence relations for logical deductive systems are reflexive, transitive and monotonic.

---

6 Some Extended Examples

6.1 Introduction

In Chapter 3 we introduced the fundamental notion of interpretation and said what it meant for an interpretation to be a model for a theory in a formal language. It should be clear that there is a use for interpretations in computer science. Either explicitly, when new programming languages are designed, or less explicitly when we invent data structures in solving some computational problem, we set up a relationship between parts of the language and the objects that we want to reason about. In both cases we can use the interpretation and the proof theory of the logic to state and derive properties of the programming language or the data structures.

6.2 Theory examples

In the next few sections we will give examples of theories and models. For this presentation our reasoning will be largely informal; that is, we will not be working within the formal systems given in Chapters 2 and 3, though we will, of course, be respecting the meanings given there to the constants and other symbols that we introduced.

6.2.1 A simple theory

Consider a language with one predicate symbol P and no function symbols or names, so in the notation of Chapter 3 we have $L(\langle\{P\},\{\ \},\{\ \}\rangle)$. Let $\mathscr{P}$ be the theory generated by the axioms

$$(\mathrm{P}_1)\quad \forall x\forall y\forall z((P(x,y)\wedge P(y,z))\rightarrow P(x,z))$$

$$(\mathrm{P}_2)\quad \forall x(\neg P(x,x))$$

To find models of $\mathrm{Th}(\{\mathrm{P}_1,\mathrm{P}_2\})$ we need only look for models of P_1 and P_2, by definition. One such model would be the interpretation whose universe is the set $\mathbf{Z}$ of positive and negative integers, and where P denotes the 'less-than' relation over $\mathbf{Z}$.

It can get tedious using the strictly formal language when we have a particular intepretation in mind, so we will allow ourselves to write

$$(\mathrm{P}_1)\quad \forall x\forall y\forall z((x\prec y\wedge y\prec z)\rightarrow x\prec z)$$

$$(\mathrm{P}_2)\quad \forall x(\neg x\prec x)$$

but we must remember that this is still a formal system and so we are still at liberty to take interpretations for these sentences that are 'non-standard', that is in which $\prec$ has some unusual denotation.

In one of the standard interpretations, where $\prec$ is the usual arithmetic 'less-than' relation over $\mathbf{Z}$, it is clear that many other sentences in addition to P_1 and P_2 are true. For example

$$\forall x\forall y(x\prec y\rightarrow\neg y\prec x)$$

but is this true in every model of P_1 and P_2?

We can reason informally to start with. Suppose there were elements in the universe (where we invent some names a and b to denote them) such that both $a\prec b$ and $b\prec a$ were the case. Then by P_1 we would have $a\prec a$, which contradicts P_2. So it cannot be the case that such elements exist, and it would be the case that

$$\neg\exists x\exists y(x\prec y\wedge y\prec x)$$

which is equivalent to

$$\forall x\forall y\neg(x\prec y\wedge y\prec x)$$

which is equivalent to

$$\forall x\forall y(\neg x\prec y\vee\neg y\prec x)$$

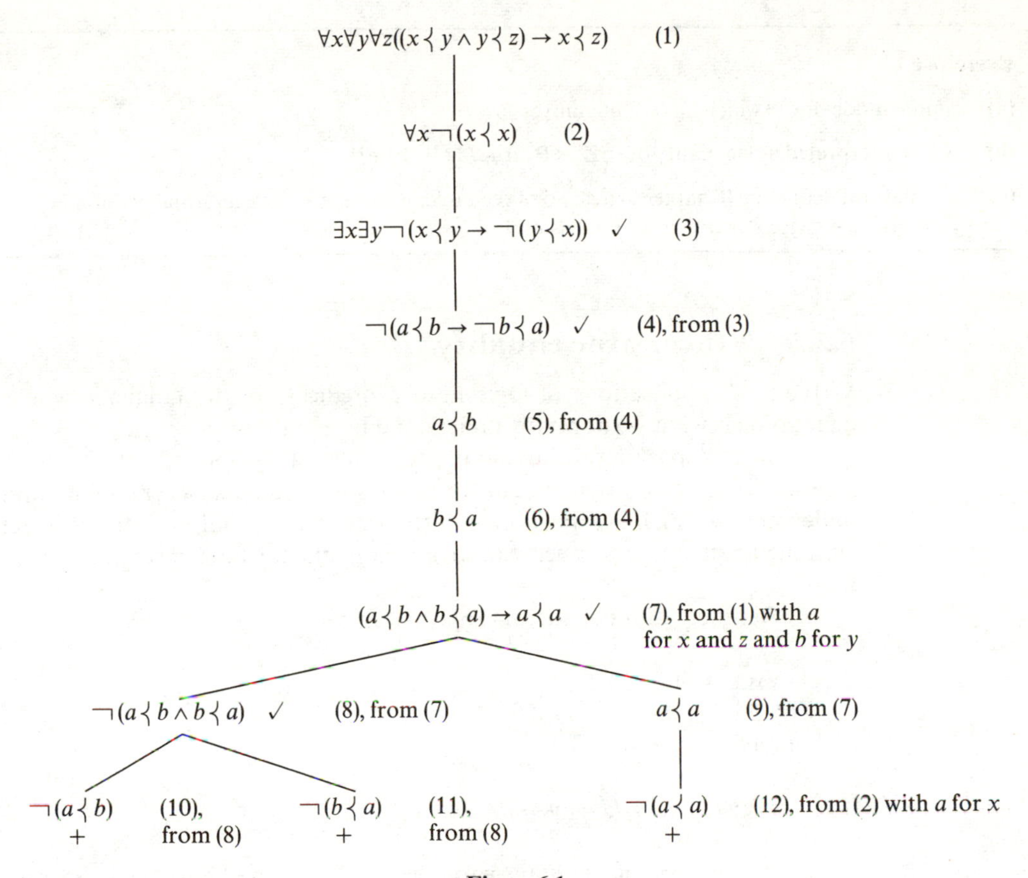

Figure 6.1.

which is equivalent to

$$\forall x \forall y(x \prec y \rightarrow \neg y \prec x)$$

which is the sentence we suggested. So, this sentence is true in every model for P_1 and P_2.

To give a formal proof of the result, we can, for example, use the tableau method. As usual, in setting up the proof, the initial sentences are P_1 and P_2 together with the negation of the sentence that we wish to prove. The tableau we get is given in Figure 6.1.

In examining this tableau you should bear in mind that the algorithms given in Chapter 4 are not, in general, going to give the shortest proof, and an algorithm which orders use of the rules differently will give a different proof. When you have studied the tableau which solves this problem you should try using the rules in different orders, and also apply your intelligence, to get different, and perhaps better or less voluminous proofs.

Exercise 6.1

(a) Find a model for $\mathscr{P}$ which has a finite universe.

(b) If P is interpreted as inequality over **Z**, is P_2 true? Is P_1 true?

(c) Use natural deduction (Chapter 5) instead of the tableau method to give a formal proof that $\forall x \forall y(x \prec y \rightarrow \neg y \prec x)$ is true in all models of $\mathscr{P}$.

6.2.2 A theory for equality

Virtually all applications of logic need a predicate in the language whose denotation is what we normally understand by equality.

Strictly speaking we should use some neutral symbol of arity 2, say Q, for this. Once again, to make our formulas and sentences easier to read and understand, we shall use the more intuitive informal symbol $=$, with the usual warning about being at liberty to give it non-standard interpretations if we wish.

Our theory for equality has one axiom

(E_1) $\forall x(x = x)$

and one axiom schema

(E_2) $\forall x \forall y \forall z_1 \ldots \forall z_n(x = y \rightarrow (\mathscr{S} \rightarrow \mathscr{S}[y/x]))$

where $\mathscr{S}$ is any formula with free variables $z_1, \ldots, z_n$, x and possibly y. $\mathscr{S}[y/x]$ is the formula that results from putting y instead of one or more of the free occurrences of x. We need a schema like this since we do not know in advance what other predicate symbols will be in the language in addition to equality.

In particular we can use $=$ as a predicate symbol in $\mathscr{S}$ and consider $\mathscr{S}$ to be the formula $x = z_1$. Then an instance of E_2 would be

$$\forall x \forall y \forall z_1(x = y \rightarrow (x = z_1 \rightarrow y = z_1)) \tag{6.1}$$

We can now use this together with E_1 to prove other useful properties of $=$. For example we can prove that $=$ is symmetric. Our informal reasoning is as follows. By the above we have

$$(a = b \rightarrow (a = a \rightarrow b = a)) \tag{6.2}$$

with a for x and z_1 and b for y. Now assume that

$$a = b \tag{6.3}$$

then, by modus ponens on 6.2 and 6.3, we have that

$$a = a \rightarrow b = a \tag{6.4}$$

but, by E_1, we have

$$a = a \tag{6.5}$$

and so, by using modus ponens again on 6.4 and 6.5, we have

$$b = a$$

We have therefore shown that, if $a = b$ is the case, then $b = a$ is the case. So we have shown (see the beginning of Chapter 5) that

$$a = b \rightarrow b = a$$

and since a and b were arbitrary it follows that we have

$$\forall x \forall y(x = y \rightarrow y = x)$$

which is the formal language statement of the symmetry of = as required.

It should be apparent that the informal demonstration we have just given is very similar in style to the informal argument of Chapter 5 (Section 5.1.3) that was there the starting point for a formal proof using natural deduction. You should carry out a similar exercise and construct the formal proof by natural deduction that corresponds to the informal argument above. Again, we give the tableau proof in Figure 6.2.

Exercise 6.2

(a) Show that transitivity of equality follows from E_1 and an instance of E_2.

(b) Give natural deduction proofs of the symmetry and transitivity of equality.

6.2.3 A theory for strings

In this case the intended interpretation is to be strings over a given alphabet of symbols, say $\{\alpha, \beta, \gamma\}$, so that, for example, $\alpha\beta$, $\alpha\alpha$, γ, $\beta\alpha\alpha\gamma$ are in the universe, as is the empty string which we shall denote by ε.

The formal language has the usual set of variables, connectives and punctuation plus a constant symbol, or name, e which denotes the empty string. The language has an infixed function symbol '.' denoting the operation of attaching a single character on the left-hand end of a string, and two

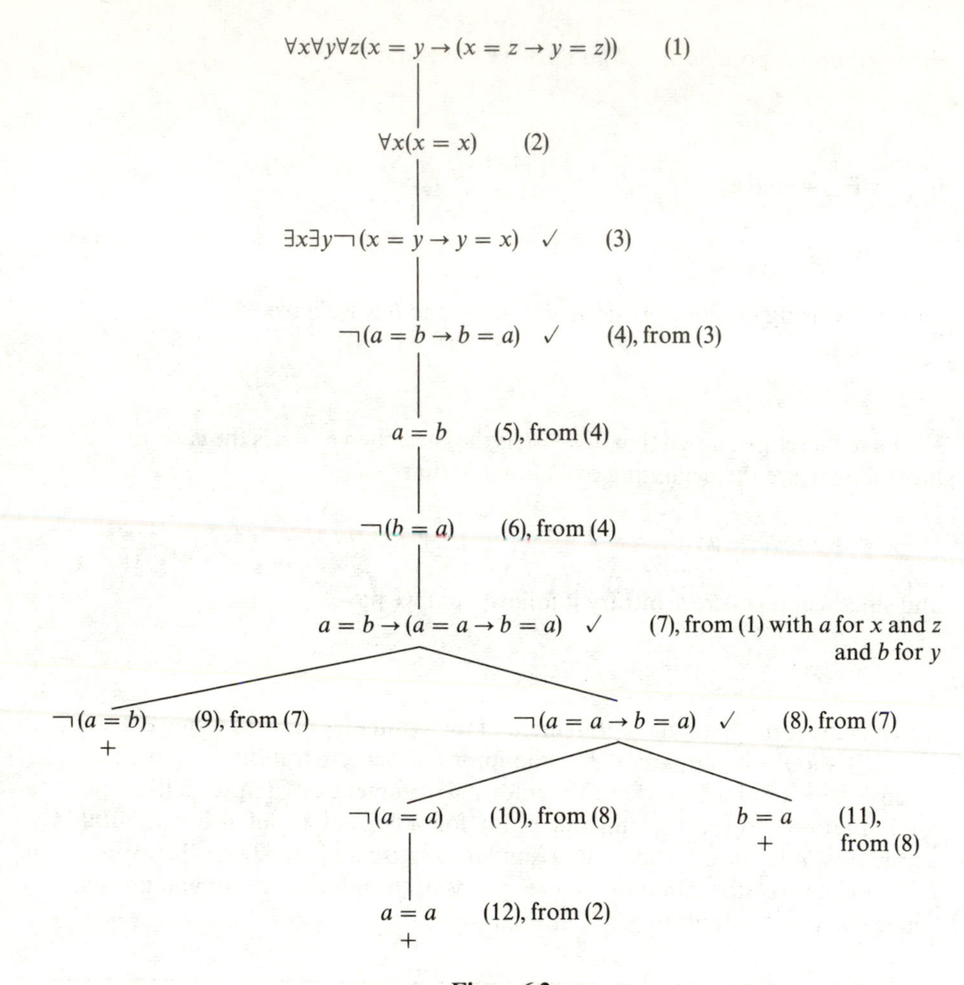

Figure 6.2.

predicate symbols C and S, $C(x)$ being true if the denotation of x is a single-character string, and $S(x)$ being true if x denotes a string.

Other names in the language are a, b and c (whose intended denotations are the single-character strings α, β and γ). In the notation of Chapter 3 the language is $L(\langle\{C, S\}, \{.\}, \{a, b, c, e\}\rangle)$. Terms in the language therefore include $a.(b.e)$, b, e (but not for example ab). Different terms may denote the same string. The denotation of some terms may be undefined.

Now we consider the theory generated by the equality axioms E_1 and E_2 plus

(S_1) $S(e)$

(S_2) $\forall x \forall y(C(x) \wedge S(y) \rightarrow S(x.y))$

(S_3) $\forall x(C(x) \rightarrow x.e = x)$

We can check to see whether some sentences are in the theory. As a first example we take

$$\forall x(C(x) \rightarrow S(x)) \tag{6.6}$$

First, assume that

$$C(a) \tag{6.7}$$

then, from S_2 we have

$$C(a) \wedge S(e) \rightarrow S(a.e) \tag{6.8}$$

with x in S_2 as a and y as e. Then, from 6.7 and S_1 we have

$$C(a) \wedge S(e)$$

and this, together with 6.8 gives

$$S(a.e) \tag{6.9}$$

Then, S_3, with x as a gives

$$C(a) \rightarrow a.e = a$$

and this together with 6.7 again gives

$$a.e = a$$

and this further, by properties of identity together with 6.9 gives

$$S(a)$$

Then, since a was arbitrary, we have the sentence 6.6 as required. So, we have shown (informally of course) that

$$\{S_1, S_2, S_3, E_1, E_2\} \vdash (1)$$

which is just to say that (1) is in the theory $Th(\{S_1, S_2, S_3, E_1, E_2\})$ since we also have, by the soundness theorem

$$\{S_1, S_2, S_3, E_1, E_2\} \models (1)$$

Once again this informal argument can be a starting point for developing a

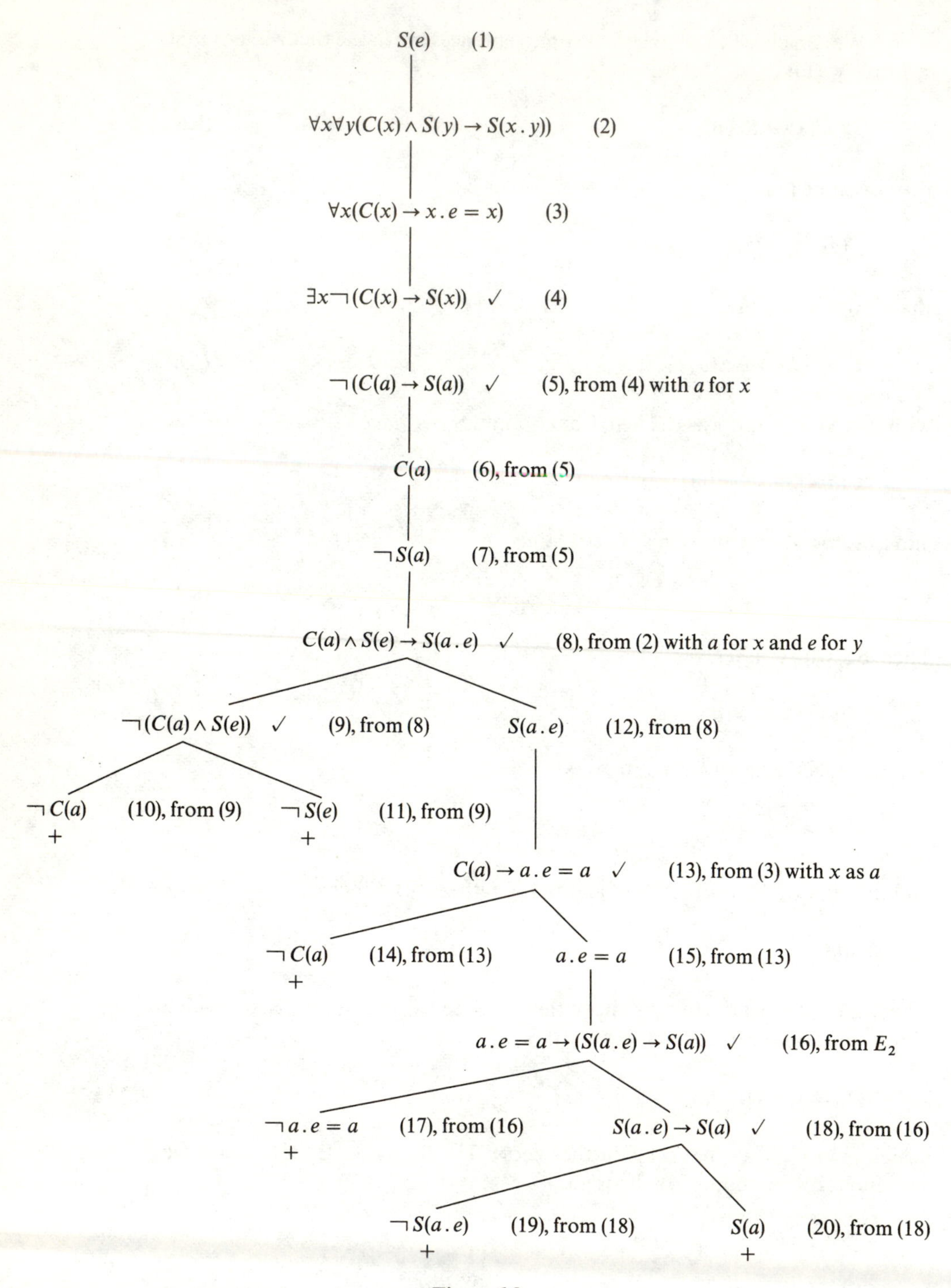
$S(e)$ (1)
$\forall x \forall y(C(x) \wedge S(y) \rightarrow S(x.y))$ (2)
$\forall x(C(x) \rightarrow x.e = x)$ (3)
$\exists x \neg(C(x) \rightarrow S(x))$ ✓ (4)
$\neg(C(a) \rightarrow S(a))$ ✓ (5), from (4) with a for x
$C(a)$ (6), from (5)
$\neg S(a)$ (7), from (5)
$C(a) \wedge S(e) \rightarrow S(a.e)$ ✓ (8), from (2) with a for x and e for y
$\neg(C(a) \wedge S(e))$ ✓ (9), from (8)
$S(a.e)$ (12), from (8)
$\neg C(a)$ (10), from (9)
+
$\neg S(e)$ (11), from (9)
+
$C(a) \rightarrow a.e = a$ ✓ (13), from (3) with x as a
$\neg C(a)$ (14), from (13)
+
$a.e = a$ (15), from (13)
$a.e = a \rightarrow (S(a.e) \rightarrow S(a))$ ✓ (16), from E_2
$\neg a.e = a$ (17), from (16)
+
$S(a.e) \rightarrow S(a)$ ✓ (18), from (16)
$\neg S(a.e)$ (19), from (18)
+
$S(a)$ (20), from (18)
+

Figure 6.3.

formal natural deduction proof that the sentence 6.6 can be derived from S_1, S_2, S_3, E_1 and E_2.

The tableau that proves this is given in Figure 6.3.

6.2.4 Another example

Consider the theory given by the equality axioms E_1 and E_2 together with

(St_1) $\forall x \forall y(f(g(x, y)) = x)$

(St_2) $N(e) \wedge \forall y(\neg e = y \leftrightarrow \neg N(y))$

(St_3) $\forall x \forall y(\neg N(g(x, y))$

We will call this set of sentences *St*. Then, as usual, of all the possible interpretations of the non-logical symbols N, e, f, g and $=$, only certain subclasses of them form models of *St*. Therefore, only a subset of all possible sentences in the language are in the theory determined by these models, that is only a subset of all possible sentences are theorems.

Clearly, we only talk in a non-trivial way if our sentences 'pick out' a subset of all possible relations that hold in the world. If the only sentences that we could utter were all valid sentences then, since they would be, by definition, true in all interpretations, we would not be saying anything interesting about particular interpretations.

Thinking back to Chapters 2 and 3, we used the word 'axiom' in defining formal deductive systems. There the axioms were all valid. How can it be, considering what has just been said, that the formal deductive system is saying anything interesting? Well, the point is that those axioms are only valid under our underlying assumption that all statements are either true or false. Without that assumption those axioms are no longer necessarily valid. To put it another way, by framing the definitions so that those axioms are valid we are selecting interpretations in which all sentences are either true or false as a framework for our logic. So, the axioms are telling us something after all.

Going back to *St*, let us fix our interpretation of the non-logical symbols as one which is a model for St_1, St_2, St_3, E_1 and E_2, and call the model M. Now consider the sentence

$$\neg N(g(a, g(b, e)))$$

Is this true or false in M? We can take axiom St_3 above and substitute a for x and $g(b, e)$ for y to get the required sentence. Therefore, since this is an instance of St_3 (using a version of P5 from Chapter 3, Section 3.2) we have established that the sentence is a theorem. You may have already noticed that *St* is an approximation to a theory of stacks (a stack is a data structure that can only be modified in a 'last in, first out' manner; you can only access the last item to have been added to it).

Exercise 6.3 Give similar arguments to decide whether or not the following are theorems of the theory St.

(a) $\neg e = g(a, e)$

(b) $f(f(g(a, e))) = a$

(c) $\forall x(N(x) \to x = e)$

6.2.5 Induction

Many proofs in mathematics and computer science require the principle of mathematical induction. This is almost invariably the case with recursively defined programs and data structures. To illustrate the formulation of induction in a logical theory we return to the theory of strings that was given in Section 6.2.3 and extend it with further axioms for another function symbol '@' whose intended interpretation is concatenation of strings. (In fact the symbol '@' is the same as the one used to denote concatenation in the SML system.)

We are going to prove by induction that '@' is associative, but before doing so, because there are minor complications in the theory of strings to do with having more than one sort of object, namely both characters and strings, we start by looking at how induction would be formulated in a simpler and more familiar setting.

If you were constructing a theory of non-negative integers you would have a predicate symbol such as I whose denotation is the integers, and axioms such as

$$I(0)$$

$$\forall x(I(x) \to I(s(x))$$

where the intended denotation of the name '0' is the number zero, and that of the function symbol 's' is the successor function over the integers, that is the function that takes an integer and adds unity to it. So, for example, $s(s(0))$ denotes the number two.

In this theory the principle of induction would be formulated as the schema

$$(\mathscr{F}[0] \wedge \forall x(\mathscr{F}[x] \to \mathscr{F}[s(x)])) \to \forall x \mathscr{F}[x]$$

where $\mathscr{F}[x]$ is any formula containing x as a free variable. In the same way as we did for equality, we need a schema for the induction axiom rather than a sentence because we want to leave open the possibility of using the principle for any statement about the domain (sentence in the formal language).

Now if we were to extend our theory of the integers with additional function symbols and axioms intended to formalize, say, the addition function, then we could use the induction schema given above to derive, for example, a sentence whose denotation is the associativity of addition.

In the theory of strings the role of the successor function is played by the denotation of the '.' function that extends a string by adding a single character to it. Recall that if a denotes α, b denotes β, c denotes γ, and so on, then $b.(c.e)$ denotes the string $\beta\gamma$ and $a.(b.(c.e))$ denotes the string $\alpha\beta\gamma$. The corresponding induction principle is an example of what is called **structural induction** – instead of arguing by induction over the integers we operate, in this case over strings, in general over formulas, trees, or indeed anything whose structure can be given the appropriate recursive definition. In fact, many of the meta-theorems proved earlier in the book have employed structural induction, but in the observer's rather than the object language.

For the theory of strings the appropriate induction schema is

$$(\mathscr{F}[e] \wedge \forall x(S(x) \rightarrow (\mathscr{F}[x] \rightarrow \forall u(C(u) \rightarrow \mathscr{F}[u.x])))) \rightarrow \forall x(S(x) \rightarrow \mathscr{F}[x])$$

This looks more complicated than it is. It says that if the formula $\mathscr{F}$ is true of the empty string and if, for all strings x, $\mathscr{F}$ is true of x implies that $\mathscr{F}$ is true of $u.x$ for any character u, then $\mathscr{F}$ is true of all strings. Because we have two sorts of object in the universe we need the $S(x)$ and $C(u)$ to ensure that the universal quantifiers can only range over the appropriate type. (There are 'many-sorted' logics in which this can be done more neatly but to discuss them would take us too far afield.)

The additional axioms for '@' (written, like the function symbol '.' that we already have, in infixed mode) are

(S_4) $\quad \forall x(S(x) \rightarrow e @ x = x)$

(S_5) $\quad \forall x \forall y \forall z(C(x) \wedge S(y) \wedge S(z) \rightarrow (x.y) @ z = x.(y @ z))$

If the term $a.(b.e)$ denotes the string $\alpha\beta$ and $b.(c.e)$ denotes the string $\beta\gamma$, then the denotation of the term $(a.(b.e)) @ (b.(c.e))$ is the string $\alpha\beta\beta\gamma$. It is clear from one's intuition of the domain that @ is associative, in other words $(t_1 @ t_2) @ t_3$ denotes the same string as $t_1 @ (t_2 @ t_3)$ whatever strings the terms t_1, t_2 and t_3 denote. To prove this we take $\mathscr{F}[x]$ in the induction schema as

$$\forall y \forall z(S(y) \wedge S(z) \rightarrow (x @ y) @ z = x @ (y @ z)) \qquad (\mathscr{F}[x])$$

We are also going to need some theorems from the theory of equality. We have already shown

$$\forall x \forall y(x = y \rightarrow y = x) \qquad \text{(symmetry of =)}$$

Using suitable instances of E_2 it is also straightforward to show

$$\forall x \forall y \forall z (x = y \rightarrow (y = z \rightarrow x = z)) \qquad \text{(transitivity of =)}$$

and, for any terms $t, t', t_1, \ldots, t_n$ and function symbol f, that

$$t = t' \rightarrow f(t_1, \ldots, t, \ldots, t_n) = f(t_1, \ldots, t', \ldots, t_n) \qquad \text{(substitution in terms)}$$

Now we take, as the inductive hypothesis, $\mathscr{F}[q]$ for some string q, so we have $S(q)$ and

$$\forall y \forall z ((q \mathbin{@} y) \mathbin{@} z = q \mathbin{@} (y \mathbin{@} z))$$

Then by $\forall$ elimination, and assuming $S(r)$ and $S(s)$, we have

$$(q \mathbin{@} r) \mathbin{@} s = q \mathbin{@} (r \mathbin{@} s) \tag{6.10}$$

Now by S_5, and the rule for substitution in terms, if p is such that $C(p)$ holds

$$((p \,.\, q) \mathbin{@} r) \mathbin{@} s = (p \,.\, (q \mathbin{@} r)) \mathbin{@} s \tag{6.11}$$

and again by S_5

$$(p \,.\, (q \mathbin{@} r)) \mathbin{@} s = p \,.\, ((q \mathbin{@} r) \mathbin{@} s) \tag{6.12}$$

From the equation 6.10, E_1, and the rule for substitution in terms, we have

$$p \,.\, ((q \mathbin{@} r) \mathbin{@} s) = p \,.\, (q \mathbin{@} (r \mathbin{@} s)) \tag{6.13}$$

From S_5 and the rule for symmetry of $=$ we have

$$p \,.\, (q \mathbin{@} (r \mathbin{@} s)) = (p \,.\, q) \mathbin{@} (r \mathbin{@} s) \tag{6.14}$$

So by several applications of the transitivity of $=$ we have, from equations 6.11–6.14,

$$((p \,.\, q) \mathbin{@} r) \mathbin{@} s = (p \,.\, q) \mathbin{@} (r \mathbin{@} s)$$

Discharging $S(r)$ and $S(s)$, and using $\forall$ introduction and the definition of $\mathscr{F}[x]$, we have $\mathscr{F}[p \,.\, q]$, and discharging $C(p)$ and using $\forall$ introduction we have

$$\forall u (C(u) \rightarrow \mathscr{F}[u \,.\, q])$$

Discharging the inductive hypothesis $\mathscr{F}[q]$ we have

$$\mathscr{F}[q] \rightarrow \forall u(C(u) \rightarrow \mathscr{F}[u.q])$$

and discharging $S(q)$ and using $\forall$ introduction we have

$$\forall x(S(x) \rightarrow (\mathscr{F}[x] \rightarrow \forall u(C(u) \rightarrow \mathscr{F}[u.x]))) \quad \textbf{(6.15)}$$

Now from S_1 and S_4 we have both

$$e @ q = q \quad \textbf{(6.16)}$$

and

$$e @ (q @ r) = (q @ r) \quad \textbf{(6.17)}$$

and so, from the equations 6.16 and 6.17, and the symmetry and transitivity of $=$, we have

$$(e @ q) @ r = e @ (q @ r)$$

But q and r are arbitrary strings, so discharging $S(q)$ and $S(r)$ and using $\forall$ introduction, we have

$$\forall y \forall z(S(y) \wedge S(z) \rightarrow (e @ y) @ z = e @ (y @ z))$$

But this is $\mathscr{F}[e]$, and so with the equation 6.15 and the principle of induction we have $\forall x \mathscr{F}[x]$, that is

$$\forall x \forall y \forall z((x @ y) @ z = x @ (y @ z))$$

which is the associativity of '@' that we set out to derive.

This use of structural induction is typical of many proofs about abstract data structures that are carried out in computer science and similar theories can be constructed for lists and trees. Manna and Waldinger (1985) give numerous examples.

6.2.6 A puzzling example

Raymond Smullyan is a well-known logician, but aside from his 'serious' work he has also produced a number of more popular books about logic that are unique in that they illustrate quite difficult topics (up to and including Gödel's theorem for example) by means of stories and puzzles. We borrow one of his puzzles now to illustrate some points about theories involving equality and also to show that the application of formal deductive systems is not just confined to reasoning in mathematics or computer science.

The story goes as follows. In *The Merchant of Venice* the character

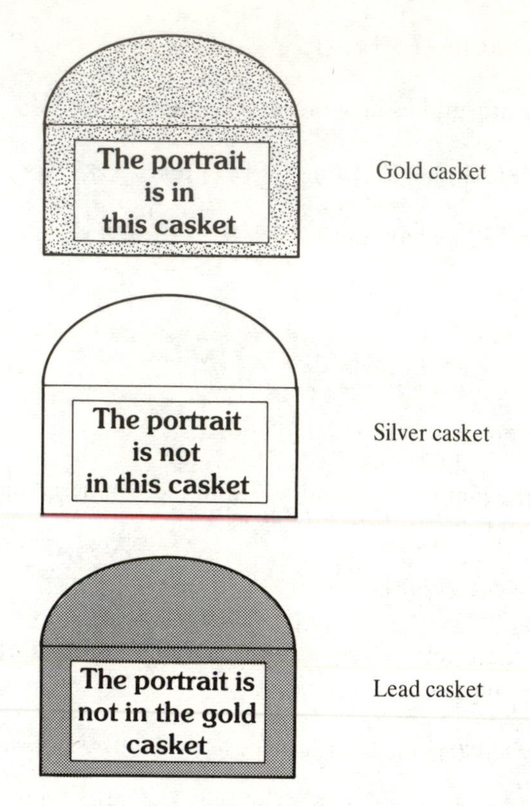

Figure 6.4.

Portia had three caskets, one made of gold, one of silver and one of lead. Concealed inside one of them was her portrait. Prospective suitors were tested by being asked which casket the portrait was in. Inscribed on the top of each casket was a statement they might find useful in making their choice. At most one of the statements is true. For example, see Figure 6.4.

It is quite easy to solve this conundrum by logical but **informal** reasoning. You should do it now as an exercise. To do it **mechanically**, either by hand or by computer, is another matter. As usual we have to set up an appropriate formal language. In this case we have three constant symbols, a, b and c say, which in the intended interpretation will denote the gold, silver and lead caskets respectively, and two predicate symbols T and P. $T(x)$ will be satisfied if the statement written on the casket that x denotes is true, $P(x)$ will be satisfied if the portrait is in the casket that x denotes. There are many such puzzles that can be constructed, but they all have one thing in common; there is only one copy of the portrait and it is in one of the caskets. So the basic theory for all these puzzles will contain a sentence that says this in the casket interpretation. There are several ways to do it. Since we know there are only three objects in the domain we can do it without quantifiers by writing

(C_1) $(P(a) \wedge \neg P(b) \wedge \neg P(c)) \vee (\neg P(a) \wedge P(b) \wedge \neg P(c)) \vee (\neg P(a) \wedge \neg P(b) \wedge P(c))$

or even without predicates by writing $P(a)$, $P(b)$ and $P(c)$ as propositional letters A, B and C.

Using quantifiers has the advantage that our theory will cope with any number of caskets, but it does require some care with the theory of equality. First we solve the puzzle, then we return to consider how to use quantifiers and the theory of equality.

The condition that at most one of the inscriptions on the caskets is true can be expressed similarly by

(C_2) $(T(a) \wedge \neg T(b) \wedge \neg T(c)) \vee (\neg T(a) \wedge T(b) \wedge \neg T(c)) \vee (\neg T(a) \wedge \neg T(b) \wedge T(c)) \vee (\neg T(a) \wedge \neg T(b) \wedge \neg T(c))$

and the inscriptions themselves by

(C_3) $T(a) \leftrightarrow P(a)$

(C_4) $T(b) \leftrightarrow \neg P(b)$

(C_5) $T(c) \leftrightarrow \neg P(a)$

Now we do not know the answer to the puzzle so we do not know exactly what we are trying to prove but we do know that it will be $P(a)$, $P(b)$ or $P(c)$. We shall opt to use the tableau method – appropriately, since Smullyan (1968) has given a particularly clear exposition of it. When the tableau is completed we shall see what is required to close it. In fact it is a straightforward exercise in the use of the $\wedge$, $\vee$ and $\leftrightarrow$ rules that we leave to the reader. All branches except one close. The open branch contains $\neg P(a)$, $P(b)$, $\neg P(c)$, showing that the silver casket is the one containing the portrait, because if we had known this at the start we would have added $\neg P(b)$ to the initial set of sentences $C_1, \ldots, C_5$ and the tableau would have closed.

If we had decided to use quantifiers to express the two conditions that the portrait is in precisely one of the boxes and no more than one of the inscriptions is true then there are several ways to say these things in the formal language. There is, however, no way to do it without using equality. Perhaps the simplest is to use the sentence

$$\exists x \exists y (\neg(x = y) \wedge T(x) \wedge T(y))$$

which says that at least two objects in the domain possess property T. We can then negate this to give the condition that no more than one of the inscriptions is true.

(C_2') $\neg \exists x \exists y (\neg(x = y) \wedge T(x) \wedge T(y))$

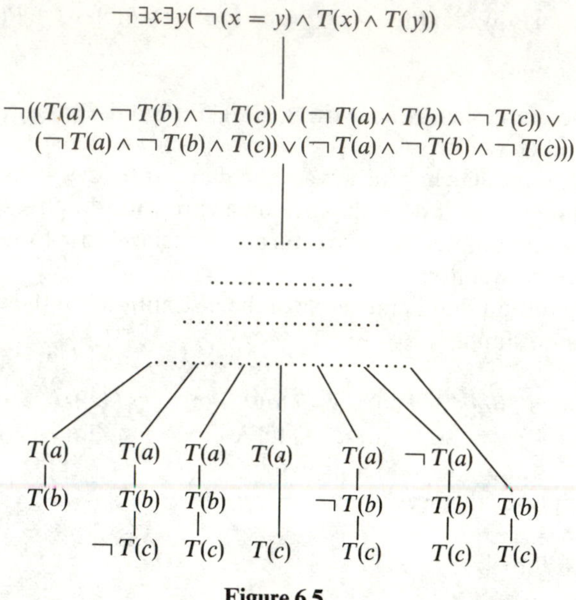

Figure 6.5.

The condition that the portrait is in precisely one of the boxes is similar, but with the additional condition that says it is in at least one of the boxes,

$$(C_1') \quad \neg\exists x\exists y(\neg(x = y) \wedge P(x) \wedge P(y)) \wedge \exists z P(z)$$

Showing that this formulation is equivalent to the one without quantifiers is instructive because it demonstrates some of the points that arise in working with theories that contain the theory of equality. We show that C_2' entails C_2. Negating C_2 and applying the tableau rules gives $3^4 = 81$ branches of which all but seven close or are duplicated. The seven distinct open branches contain (reading them vertically) the formulas shown in Figure 6.5. Note that they enumerate all possible countermodels to C_2, that is those in which two or more of the statements on the caskets are true.

Now we can close each of these branches by applying the tableau rules to the topmost sentence. First we get $\forall x\forall y\neg(\neg(x = y) \wedge T(x) \wedge T(y))$. Then taking for example a for x and b for y we get

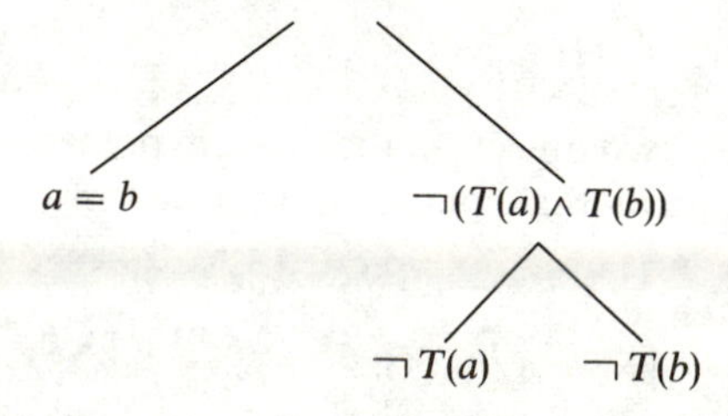

Now the two rightmost branches can be used to close the three leftmost branches of the big tableau that contain $T(a)$ and $T(b)$. The branch containing $a = b$ can be closed if our theory of equality contains sentences such as $\neg(a = b)$, $\neg(b = c)$ and $\neg(c = a)$ that say that the constants of the language denote different objects in the interpretation. The other branches of the tableau can be closed in a similar way by taking a different choice of constants, remembering that $\forall$ sentences can be used more than once.

C_1 can be dealt with in a similar way with the additional complication that it involves $\exists z P(z)$. Here another aspect of reasoning with equality comes in. Recall that $\{P(a) \vee P(b) \vee P(c)\} \models \exists z P(z)$ because we can construct the closed tableau

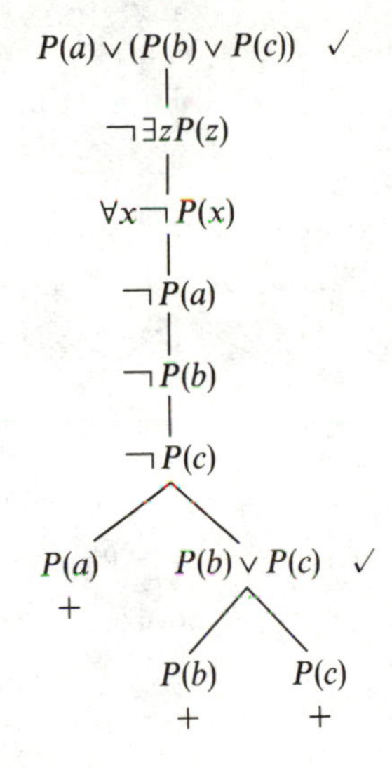

However, $\exists z P(z)$ does not entail $P(a) \vee P(b) \vee P(c)$ unless we add an extra condition that says there are no objects other than a, b and c in the domain. The appropriate condition to add is

$$\forall x((x = a) \vee (x = b) \vee (x = c))$$

This is known as the **closed-world assumption** (CWA). Together with the assumption that distinct names denote distinct objects in the domain, it is of particular importance in deductive databases and artificial intelligence where one of the aims is to write logically based programs that, for example, will enable a computer-controlled robot to prove simple theorems about its environment. With the CWA we can get the equivalence we need as shown in Figure 6.6. Now we can close all these branches by using sentence E_2 from the theory of equality, which gives for example $(d = a) \rightarrow (P(d) \rightarrow P(a))$, which when expanded closes the left-hand branch. The others are similar.

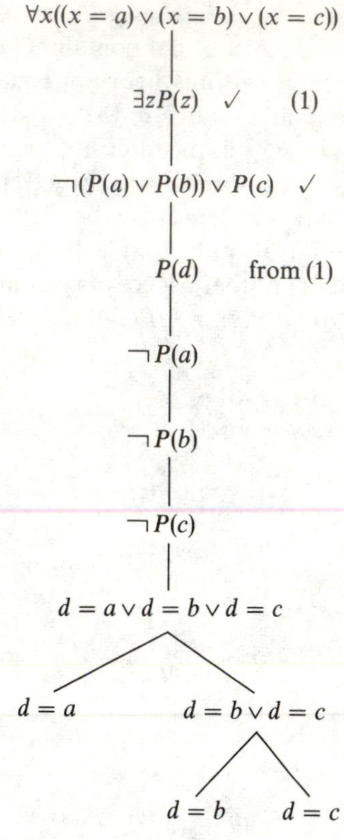

Figure 6.6.

6.3 Gödel and the limits of formalization

Because so many good accounts are available elsewhere we have chosen not to describe the classical logical theories of arithmetic and the foundations of set theory. A theory of arithmetic (more precisely – a theory whose standard interpretation is the integers and the usual arithmetic operators and predicates over them) can be generated from the theory of equality, six axioms for the arithmetic constants and operators and an induction schema. If Hilbert's assumption had been correct (see the end of Chapter 3) then all the truths of arithmetic could in principle have been established by the methods of the current chapter. However, in 1931 Gödel showed that, for a certain sentence G, neither G nor $\neg G$ can be proved within the formal system for arithmetic.

The method Gödel used was to show that formulas and proofs could be coded as numbers (Gödel numbers), so that statements in the observer's language about the existence of proofs in the formal system can be expressed as statements about numbers within the system itself. The formula G is a rather unusual one, in that it can be interpreted as asserting its own unprovability. Nevertheless, its denotation can clearly be recognized as true, which means that there are truths of arithmetic that cannot be proved or disproved within the formal system for arithmetic.

Furthermore, Gödel showed that no formal system for arithmetic that is consistent can be complete. It might be thought that all one needs to do is to add all the truths of arithmetic that cannot be proved within the system as extra axioms. The catch here is that the set of Gödel numbers of the sentences that are true in arithmetic is not effectively computable, so once again we have come up against one of the fundamental limits to computation. All this meant that Hilbert's programme for the formalization of mathematics could never be realized in its entirety, although it must be said that the attempt to do so had led to several of the most significant discoveries in the history of mathematics.

SUMMARY

- A *theory* is a set of sentences, all of which are true in a class of interpretations. A theory can be specified by giving the *axioms* from which all the other sentences in the theory can be deduced by one of the systems of deduction for predicate calculus. The soundness of the deduction system guarantees that, if the axioms are true in the intended class of interpretations, then all the sentences of the theory are true in those interpretations. Theories provide the means by which logic is actually used to formalize or automate reasoning in other domains such as mathematics, computer science and artificial intelligence.
- Theories of *ordering* and *equality* are often required as part of other theories. Axiomatizations have been given for both these theories and some theorems have been deduced from them. With such commonly used theories we often relax the rules of the logical language slightly, allowing infixed predicate symbols to make formulas easier to read.
- Strings and stacks are examples of *abstract data structures*. Axiomatizations have been given in this chapter and some theorems of these theories have been deduced.
- Many proofs in mathematics and computer science use the principle of mathematical induction. This can be formalized by including in the theory an *induction schema*. An appropriate induction schema has been given for the theory

of strings and used to prove the associativity of string concatenation.

- Most theories have to include the theory of equality (identity) and, particularly in applications of logic in artificial intelligence, often require further axioms of identity stating, for example, that different names denote different objects, and the closed world assumption which says that no objects exist other than those named. We have illustrated these aspects of automating practical reasoning by using logic to solve a well-known type of puzzle.

7 Logic Programming

7.1 Introduction

7.1.1 Background

We have already seen how logic can be used to reason about the objects or data structures that computer programs manipulate and we mentioned in Chapter 1 that another use for logical notation in computer science is as a specification language. Given a problem to be solved or task to be carried out, what constitutes a solution is specified in some more or less formal notation. A method or algorithm for producing the required output is then invented and coded as a sequence of instructions that can be executed to produce an output that satisfies the specification. A necessary supplementary activity is proving, possibly using some of the methods we have already discussed, that the output satisfies the specification.

The idea behind logic programming is that, having written down a specification of what constitutes a solution to a problem in some logical language, the execution mechanism or operational semantics is then given by a system of deduction for this language. Not only are the steps of inventing and coding an effective algorithm saved but, since a provably sound deduction mechanism is used, showing additionally that the output satisfies the specification is unnecessary.

Potentially many formulations of logic programming are possible, depending on the logic and deduction mechanism used. A common mistake is to confuse logic programming with programming in Prolog, which is just one particular (and impure) logic programming system. For readers with the background to make sense of the remark Prolog bears much the same relationship to logic programming as Lisp does to the λ-calculus. However, while the core of Lisp is a complete implementation of the λ-calculus, the purely logical core of Prolog implements only a subset of the full predicate calculus and is supplemented, in the systems used in practice, by non-classical negation and other meta-logical constructions. There is no theoretical reason why the idea of using logic as a programming language should not be extended to the full predicate calculus but, at the present state of development, implementations are not efficient enough to be attractive for most programming tasks. With its non-logical extensions Prolog has the full power of any other language and yet is sufficiently close to the ideal of programming in logic to have generated an explosion of interest in the possibilities. As we described at the end of Chapter 3, Church and Turing showed that logic and the λ-calculus have equivalent computing power in the sense that the same set of functions is computable in both formalisms. This does not address the possibility that formulating algorithms may be more efficiently carried out in one formalism than the other. In Chapter 3 we also briefly mentioned the notion of one formalism being of **higher order** than another, and the λ-calculus is of higher order than predicate calculus in just this sense. Whether this makes programming systems based on the λ-calculus better in any theoretical or practical sense than those based on logic programming is still a matter of current debate.

Prolog was developed around 1972 by Colmerauer and his group in Marseilles as a tool for processing natural language. Kowalski, who had some contact with the group, realized that there are two ways of looking at a Prolog program. Either you can understand it procedurally, as you would a programming language such as Pascal, or you can read it as a set of sentences in (a subset of) predicate calculus. Of course it is not just logic programming languages that have this characteristic. Functional languages such as SML can also be seen, partly at least, as executable specifications. As we have indicated, there is some controversy among computer scientists over whether programming with functions is more or less effective than programming with relations. The arguments centre round notions such as higher order functions, that is functions which take other functions as arguments or return other functions as results, and logical variables, that is variables in the sense that we have been using them in logic rather than in the way that they are used in languages such as Pascal. For the straightforward tasks in this book there is very little to choose between the two systems.

7.1.2 Definite clauses

In this book we look at only one variety of logic programming, a subset of predicate calculus called definite clause programs. In order to make the analogy with programming as close as possible it is customary to change the logical notation slightly and introduce a new way of writing formulas and some new terminology.

> **Definition 7.1:** A **definite clause** is a formula either of the form $P \leftarrow$ or of the form $P \leftarrow Q_1, Q_2, \ldots, Q_n$, for $n > 0$, where P and the Q_i are atomic formulas of predicate calculus, that is of the form $S(t_1, \ldots, t_k)$ where S is a predicate symbol and the t_i are terms. In both cases P is said to be the **head** of the clause (we will draw an analogy between this and a procedure heading in, say, Pascal) and $Q_1, Q_2, \ldots, Q_n$ is the **body** of the clause.

> **Definition 7.2:** A **definite clause program** is a sequence of definite clauses.

For example

$P(x) \leftarrow Q(x, y), R(y)$
$Q(a, b) \leftarrow$
$R(b) \leftarrow$

is a definite clause program.

Logically speaking, the definite clause $P \leftarrow Q_1, Q_2, \ldots, Q_n$ is to be read as 'P if Q_1 and Q_2 and Q_2 and ... and Q_n', and semantically is exactly the same as the predicate calculus formula $Q_1 \wedge Q_2 \wedge \ldots \wedge Q_n \rightarrow P$. The meaning of $P \leftarrow$, a definite clause with null body, is that P holds unconditionally, that is simply P; the meaning of $\leftarrow P$ is that there are no circumstances in which P holds, that is $\neg P$; the meaning of $\leftarrow$ by itself is $\perp$.

If P and the Q_i contain variables, then the meaning of $P \leftarrow Q_1, Q_2, \ldots, Q_n$ is the same as $\forall x_1 \ldots \forall x_j \exists y_1 \ldots \exists y_k (Q_1 \wedge Q_2 \wedge \ldots \wedge Q_n \rightarrow P)$ where $x_1 \ldots x_j$ are the variable symbols in P and $y_1 \ldots y_k$ are the variable symbols that occur in the Q_i but not in P. For example $P(x) \leftarrow Q(x, y), R(y)$ means $\forall x \exists y (Q(x, y) \wedge R(y) \rightarrow P(x))$.

A set of definite clauses is like a set of procedure definitions; to use the procedures we have to execute a procedure call. To execute a definite clause program we add to it a **goal**, which is a formula of the form $\leftarrow G_1, G_2, \ldots, G_n$. Logically what we are doing is to show that $G_1 \wedge G_2 \wedge \ldots \wedge G_n$ follows from the program. As in the tableau method the proof is by refutation, but the inference mechanism is different. We use a new rule called **resolution**.

Before going into the details of what resolution is we explore a bit further the connection with programming by looking again at one of the examples from Chapter 6. There we discussed a theory of strings, the language for which contained two function symbols, one being the '.' which denotes the 'cons' function that takes a character and a string as arguments and adjoins them. The other function symbol '@' denoted concatenation of strings. We shall use concatenation as an example of programming with definite clauses.

To do this we have to rewrite it as a relation over strings and, as is customary in programming, we give it a mnemonic symbol. The distinction between characters and strings that was preserved in Section 6.2.8 can be omitted (and in fact we then get a slightly more general theory of sequences (lists) in which the elements of sequences can be general terms, including sequences). The axioms S_4 and S_5 from Section 6.2.5 then become

$$\forall v\, \text{Concat}(e, v, v)$$
$$\forall w \forall x \forall y \forall z\, (\text{Concat}(x, y, z) \rightarrow \text{Concat}(w.x, y, w.z))$$

In the standard interpretation these sentences are easily seen to be true. The first asserts that the result of concatenating any sequence v onto the empty sequence is v. The second asserts that if the result concatenating y onto x is z then the result of concatenating y onto the sequence whose first element (head) is w and whose remainder (tail) is x, is a sequence whose head is w and whose tail is z.

In programming, this fact about sequences is used as the justification for function definitions such as

```
def concat(x,y) =
   if x = emptystring then y
   else cons(head(x),concat(tail(x),y))
endef
```

and this definition might be used by executing an instruction such as

```
print(concat("ab","c"))
```

resulting in the answer "abc" being output. The equivalent logic program consists of the two definite clauses

$$\text{Concat}(e, v, v) \leftarrow$$
$$\text{Concat}(w.x, y, w.z) \leftarrow \text{Concat}(x, y, z)$$

and these would be used by adding the goal

$$\leftarrow \text{Concat}(a.(b.e), c.e, z)$$

resulting in the answer $z = a.(b.(c.e))$ being output. To see exactly how this happens we now have to say what the resolution rule of inference is.

7.1.3 Propositional resolution

Definite clause programs are only a subset of all predicate calculus formulas and we shall only consider the simplified form of resolution that applies to definite clauses. We explain it first for the propositional case. The rule is

from	$P \leftarrow Q_1, Q_2, \ldots, Q_m$
and	$\leftarrow P, R_1, R_2, \ldots, R_n$
deduce	$\leftarrow Q_1, Q_2, \ldots, Q_m, R_1, R_2, \ldots, R_n$

It can be shown that if both the formulas above the line (the **resolvents**) are satisfiable then the formula below the line is satisfiable. So, if we can derive $\leftarrow$ from a definite clause program and a goal, then the program and the goal are together unsatisfiable. This means that the program entails the negation of the goal, that is entails the formula to the right of the $\leftarrow$ in the goal. The basic idea is the same as the tableau method, but the inference rule used is, superficially at least, entirely different. Here is an example (the numbers are not part of the program; they are there for explanation).

Program	(1)	$P \leftarrow Q, R$
	(2)	$Q \leftarrow$
	(3)	$R \leftarrow$
Goal	(4)	$\leftarrow P$
Resolution using (4) and (1)	(5)	$\leftarrow Q, R$
Resolution using (5) and (2)	(6)	$\leftarrow R$
Resolution using (6) and (3)		$\leftarrow$

We have shown that formulas (1)–(4) are unsatisfiable and hence that (1), (2) and (3) entail P, as of course is obvious from their logical meaning.

We can compare this with the tableau for the same initial sentences, that is $(Q \wedge R) \rightarrow P, Q, R$ and $\neg P$. The tableau we get, as you should check, is shown in Figure 7.1.

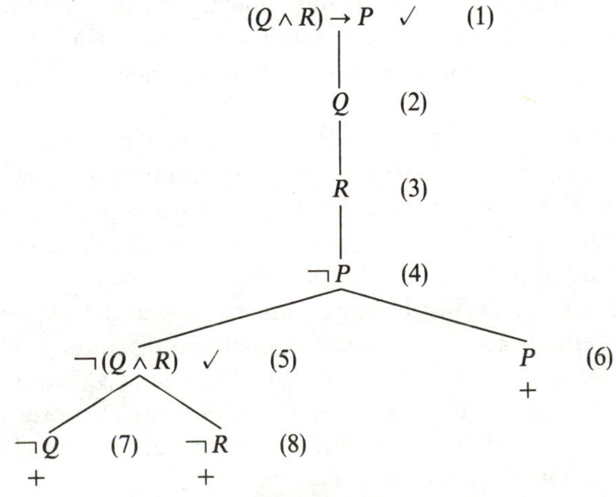

Figure 7.1.

Here (1) gives (5) and (6), and (5) gives (7) and (8). The connection with the resolution proof is made clearer if we note that (4) and (6), which came from (1), contradict. Then (7), which came from (5), contradicts (2). Finally (8), which also came from (5), contradicts (3). The same links between contradictory sentences, and the sentences that they came from, can be seen in the resolution proof. In fact, this similarity can be made even more precise, which makes resolution and tableaux look like variants of the same logical idea, which of course is exactly what they are.

For predicate calculus the resolution rule is essentially the same, but the presence of variables introduces the additional possibility of creating resolvents by making substitutions. Essentially what we do is to make atoms the same by suitable variable substitutions. To say this more precisely we will, at this point, have to go into considerable detail before we can return to logic programming. However, the ideas involved are fundamental not just to logic programming, or even to the more general subject of theorem-proving in full predicate calculus, but also to other areas of computer science such as functional programming and term re-writing. So a substantial digression at this point is worthwhile.

7.2 Substitution and unification

7.2.1 Substitution

Definition 7.3: A **substitution** is a function from variable symbols to terms in which all but a finite number of variable symbols are mapped into themselves. Substitutions therefore can be presented as a finite set of variable/term pairs $\{v_1/t_1,\ldots,v_n/t_n\}$ in which the v_i are distinct variable symbols and no t_i is the same as the corresponding v_i. A pair v_i/t_i is called a **binding** for v_i. In computer science a substitution is often called a **set of variable bindings** or an **environment**.

A substitution σ can be extended to a function σ^* from terms to terms in which a term t is mapped into the term obtained from t and σ by, for every v_i/t_i pair in σ such that v_i occurs in t, simultaneously replacing all such v_i in t by the corresponding t_i. By a mild abuse of notation the image of t under σ^* is written $t\sigma$ (rather than $t\sigma^*$) and $t\sigma$ is said to be an **instance** of t. Note that we say 'simultaneously replacing' because some t_j, or even t_i itself, may contain v_i and this occurrence is not to be subject to further replacement. This notion of a single set of simultaneous replacements can be specified more formally but only by going into some mathematical detail. For our purpose the computer programs given below make the definition completely precise. The following examples should make it intuitively clear as well.

Suppose the formal language has constants a and b, variable symbols

x, y and z, and function symbols f and g. If t is the term $f(x, g(y), b)$ and σ is $\{x/a, y/x\}$, then $t\sigma$ is $f(a, g(x), b)$. If t is the term $f(x, y, z)$ and σ is $\{x/y, y/z, z/x\}$, then $t\sigma$ is $f(y, z, x)$.

Substitutions often have to be applied one after the other, which means we need the following definition.

Definition 7.4: If σ and θ are substitutions then the **composition** $\sigma\theta$ of σ and θ satisfies $(t\sigma)\theta = t(\sigma\theta)$ for any term t.

Let σ be the substitution $\{x_1/s_1, \ldots, x_m/s_m\}$ and θ be the substitution $\{y_1/t_1, \ldots, y_n/t_n\}$. Since function symbols and constants are unchanged by substitution we need only consider variable symbols. Let $X = \{x_1, \ldots, x_m\}$ and $Y = \{y_1, \ldots, y_n\}$ and let z be any variable. There are three cases to consider. Suppose $z = x_i$ in X. Then $(z\sigma)\theta = (x_i\sigma)\theta = s_i\theta$ so, if $(z\sigma)\theta = z(\sigma\theta)$, we must have $x_i/s_i\theta$ in $\sigma\theta$. If z is also y_j in Y we must omit y_j/t_j from $\sigma\theta$. If z is not in X but is y_j in Y then, remembering that σ implicitly contains z/z, we have $(z\sigma)\theta = z\theta = y_j\theta = t_j$ so we must have y_j/t_j in $\sigma\theta$.

So $\sigma\theta$ is the substitution $\{x_1/s_1\theta, \ldots, x_m/s_m\theta, y_1/t_1, \ldots, y_n/t_n\}$ in which pairs $x_i/s_i\theta$ for which $x_i = s_i\theta$ and pairs y_j/t_j for which $y_j = x_i$ for some i have been deleted. For example, if σ is $\{x/g(y), y/z\}$ and θ is $\{x/a, y/b, z/y\}$ then $\sigma\theta$ is $\{x/g(b), z/y\}$.

Note that, if η is another substitution, then by the result above we have $z((\sigma\theta)\eta) = (z(\sigma\theta))\eta = ((z\sigma)\theta)\eta = (z\sigma)(\theta\eta) = z(\sigma(\theta\eta))$. Since this holds for any variable, $(\sigma\theta)\eta = \sigma(\theta\eta)$ and so composition of substitutions is associative. Furthermore the null substitution $\{\ \}$ is a left and right identity and so substitutions form a monoid under composition.

The following Prolog predicates define the function σ^* introduced above (and the SML versions are given in Appendix B).

```
apply_subst(Subst, Term, NewTerm):- Term=..[Symbol|Args],
   (Args=[],!,(member(Symbol/NewTerm,Subst),!; NewTerm=Term);
   do_args(Subst,Args,NewArgs), NewTerm=..[Symbol|NewArgs]).
do_args(_,[],[]).
do_args(Subst,[FirstArg|RestofArgs],[NewArg|RestofNew]):-
   apply_subst(Subst,FirstArg,NewArg),
   do_args(Subst,RestofArgs,RestofNew).
```

and the following define composition of substitutions.

```
composition([Var/Term|Sub1],Sub2,Sub12):-composition(Sub1,Sub2,S),
   delete_pair(Var,S,Sub), apply_subst(Sub2,Term,NewTerm),
   (Var=NewTerm,!,Sub12=Sub; Sub12=[Var/NewTerm|Sub]).
composition([],Sub,Sub).
delete_pair(_,[],[]).
delete_pair(X,[V/T|Y],Z):-X=V,!,Z=Y;
                              delete_pair(X,Y,U),Z=[V/T|U].
```

Exercise 7.1

(a) Say which of the following are valid substitutions, if invalid say why:
  (i) $\{x/y, y/x\}$
  (ii) $\{x/x, y/x\}$
  (iii) $\{x/y, x/z\}$
  (iv) $\{x/f(x, y), y/g(z)\}$
  (v) $\{x/a, y/x\}$
  (vi) $\{z/a, b/c\}$

(b) Apply the substitution $\{x/h(y, a), y/b, z/g(c)\}$ to each of the following terms:
  (i) $f(x, y, z)$
  (ii) $f(a, b, c)$
  (iii) $h(x, x)$
  (iv) $f(g(h(x), v), g(x, y), g(z, f(x, y, u)))$

(c) Calculate the compositions $\sigma\tau$ and $\tau\sigma$ where σ and τ are respectively
  (i) $\{x/y, y/x\}$ and $\{x/f(y), z/a\}$
  (ii) $\{x/a, y/z, z/f(x, y, a)\}$ and $\{x/b, z/g(x)\}$
  (iii) $\{x/z, z/y\}$ and $\{z/x, y/z\}$

7.2.2 Unification

Definition 7.5: A **unifier** of two expressions F and G is a substitution σ that makes $F\sigma = G\sigma$.

A single pair of expressions may have no unifier or several. For example $f(x, y)$ and $f(x, a)$ not only have $\{y/a\}$ as a unifier, but also $\{x/a, y/a\}$.

Definition 7.6: Let Σ be the set of all unifiers of F and G. A substitution $\mu \in \Sigma$ having the property that $\mu\sigma = \sigma$ for any $\sigma \in \Sigma$ is called a **most general unifier** (mgu) of F and G.

It is by no means obvious that such a substitution exists: our proof that it does will be **constructive**. If such a substitution exists the proof will actually construct the substitution. We give a method that is guaranteed either to find the mgu of two expressions or to tell us that it does not exist, in which case $\Sigma = \{\ \}$.

We construct a finite sequence of triples $\langle F_0, G_0, \mu_0\rangle, \ldots, \langle F_n, G_n, \mu_n\rangle$ such that $F_0 = F$, $G_0 = G$, $\mu_0 = \{\ \}$, $F_n = G_n$ and $\mu_n = \mu$, an mgu of F and G, where

$$F_{i+1} = F_i\sigma_i$$
$$G_{i+1} = G_i\sigma_i$$
$$\mu_{i+1} = \mu_i\sigma_i$$

and σ_i is $\{v_i/t_i\}$ where v_i is a variable symbol in F_i or G_i and t_i is a term in which v_i does not occur. This last condition is the crucial one that will ensure that the algorithm terminates. If there are n distinct variable symbols in F and G taken together then, since we are eliminating one variable at each step, the sequence will contain n elements. The next step is to show how to construct the σ_i.

Given two expressions $F = f(F_1, \ldots, F_m)$ and $G = g(G_1, \ldots, G_n)$ we define a set $d(F, G)$ called the **difference set** of F and G as follows. If $F = G$ then $d(F, G) = \{\ \}$, if $F \neq G$ but $f = g$ and $m = n$ then $d(F, G) = d(F_1, G_1) \cup \ldots \cup d(F_n, G_n)$, and if $f \neq g$ or $m \neq n$ then $d(F, G) = \{F{:}G\}$.

A difference set D is **reducible** if, for every pair $U{:}V \in D$, at least one of U and V is a variable symbol, and neither U occurs in V nor V occurs in U.

A **reducing substitution** for a difference set D is a substitution $\{v/t\}$ derived from any member $U{:}V$ of D as follows: if U is a variable symbol then let $v = U$ and $t = V$; if V is the variable (one of them must be) then let $v = V$ and $t = U$. Note that a reducible difference set D may become irreducible as a result of applying a reducing substitution σ to F and G. For example if $F = f(g(x), x)$ and $G = f(y, f(y))$ then $d(F, G) = \{y{:}g(x), x{:}f(y)\}$. A reducing substitution is $\{y/g(x)\}$ and $d(F\sigma, G\sigma) = \{x{:}f(g(x))\}$ which is irreducible. Note that choosing the other substitution does not help: we still get the irreducible set $\{y{:}g(f(y))\}$. It can be shown that unification is linear, in the sense that, if two expressions are unifiable, any choice of reducing substitutions will unify them (though possibly giving mgus that are variants – the variable symbols may be permuted), while if they are not unifiable then no choice of reducing substitution will unify them. This simplifies the programming. To unify two expressions we construct, notionally at least, the triples of the sequence defined above, generating at each step the difference set $d(F_i, G_i)$. If $d(F_i, G_i)$ is irreducible then F and G are not unifiable. If $d(F_i, G_i)$ is reducible then every element is a reducing substitution and we can take σ_i as any one of them.

For example to unify $f(x, g(a, y))$ and $f(a, g(x, z))$ we can construct the sequence as set out in the following table.

i	F_i	G_i	$d(F_i, G_i)$	μ_i
0	$f(x, g(a, y))$	$f(a, g(x, z))$	$\{x{:}a, a{:}x, y{:}z\}$	$\{\ \}$
1	$f(x, g(a, z))$	$f(a, g(x, z))$	$\{x{:}a, a{:}x\}$	$\{y/z\}$
2	$f\{a, g(a, z))$	$f(a, g(a, z))$	$\{\ \}$	$\{x/a, y/z\}$

Here is another example, this time with terms $f(g(x), g(y))$ and $f(y, g(x))$, which are not unifiable. Note that $d(F_1, G_1)$ is not reducible because the pair $x{:}g(x)$ violates the condition that the variable that is one element of the pair shall not occur in the other.

i	F_i	G_i	$d(F_i, G_i)$	μ_i
0	$f(g(x), g(y))$	$f(y, g(x))$	$\{g(x){:}y, y{:}x\}$	$\{\ \}$
1	$f(g(x), g(g(x)))$	$f(g(x), g(x))$	$\{x{:}g(x)\}$	$\{y/g(x)\}$

We have seen that termination is guaranteed but have not yet shown that we get the correct answer on termination. A detailed proof of this is too long to give in full here but we will sketch the basic idea, which involves induction on the sequence of triples defined above. We show that, for all unifiers σ of F and G, $\sigma = \mu_i\sigma$ for all i, $0 \leqslant i \leqslant n$. Clearly this is true for $i = 0$ because $\mu_0 = \{\ \}$, and if it is true for $i = n$ we have shown that $\mu_n = \mu$ is an mgu of F and G. Whatever choice we make for σ_i from $d(F_i, G_i)$ it has the form $U{:}V$ where we can suppose U is a variable not occurring in V. It is then fairly straightforward to show that $\sigma_i\sigma = \sigma$ for all unifiers σ of F and G. So $\sigma = \mu_i\sigma = \mu_i(\sigma_i\sigma) = (\mu_{i+1}\sigma)$ and hence $\mu_n = \mu$ is an mgu by the definition of mgu. It is also clear that, if the sequence terminates with non-empty difference set, then F and G are not unifiable, because, if they were, there would be a reducing substitution.

Exercise 7.2

(a) Find the difference sets for each of the following pairs of terms:
- (i) $f(a, b, c)$ and $f(x, y, z)$
- (ii) $f(a, b)$ and $g(a, b)$
- (iii) $f(f(a, b), g(c))$ and $f(f(g(a), g(b)), g(c))$
- (iv) $f(x, f(x, z))$ and $f(g(x), f(x, y))$

(b) For each of the sets that you calculated in (a) above, say whether or not they are reducible and why.

(c) Calculate, where they exist, the most general unifiers for each of the pairs of terms in (a) above. If no most general unifier exists, say why not.

(d) Under what circumstances is it true, for substitutions σ and τ, that $\sigma\tau = \tau\sigma$?

7.2.3 Some background on unification and resolution

The unification method just described, although theoretically transparent, contains a lot of redundancy in the representation and manipulation of the expressions and substitutions. For example, not only is it unnecessary to calculate F_i and G_i explicitly at each stage but the substitutions can be stored in product form rather than carrying out explicit composition. In practice much faster but less transparent algorithms are used. For more mathematical detail by far the best treatment (rigorous but very readable) of unification and resolution, and indeed of many of the basic ideas of logic, is the book *Logic:*

Form and Function by Robinson (1979). Robinson, basing his work on that of Herbrand and Prawitz, developed resolution in 1965 as the basis of automatic theorem-proving. In the ten years from 1965 to 1975 there were numerous refinements, and it was believed at the time that theorem-provers could be constructed that would be applicable to a wide variety of problems in mathematics and artificial intelligence. This early optimism has turned out to be somewhat premature and it is now considered doubtful that one single uniform procedure will ever be effective for all problems. Nevertheless, resolution is still the basis for one important subclass of automatic theorem-proving, namely logic programming, which brings us back to the main subject matter of this chapter.

7.3 Resolution

7.3.1 First-order definite clauses

The resolution rule for formulas containing variables is an elaboration, using unification, of that for the propositional case.

$$\begin{array}{ll} \text{from} & P \leftarrow Q_1, Q_2, \ldots, Q_m \\ \text{and} & \leftarrow P^*, R_1, R_2, \ldots, R_n \\ \hline \text{deduce} & \leftarrow Q_1\sigma, Q_2\sigma, \ldots, Q_m\sigma, R_1\sigma, R_2\sigma, \ldots, R_n\sigma \end{array}$$

where σ is a most general unifier of P and P^*, and the set of variable symbols occurring in $P \leftarrow Q_1, Q_2, \ldots, Q_m$ is disjoint from the set of those occurring in $\leftarrow P^*, R_1, R_2, \ldots, R_n$. If these sets are not already disjoint then they can be made so by renaming. Recall that all the variables are quantified, though we do not write the quantifier, and so they are just placeholders, and as long as we rename them uniformly, so that whenever a certain variable appears we always replace it by the same new variable, the meaning of the formula is unchanged.

Here is an example of the more general resolution rule in use:

$$\begin{array}{l} P(x, f(x)) \leftarrow Q(x, y), R(y) \\ \qquad\qquad \leftarrow P(a, z), R(z) \\ \hline \qquad\qquad \leftarrow Q(a, y), R(y), R(f(a)) \end{array}$$

Now that we have a resolution rule that handles variables we can complete the concatenate example and show how answer substitutions are calculated. The definite clause program was

(1) $\text{Concat}(e, v, v) \leftarrow$

(2) $\quad \text{Concat}(w.x, y, w.z) \leftarrow \text{Concat}(x, y, z)$

where we have numbered the program clauses for reference in explanation. To use the program to compute the result of concatenating the single element c onto the sequence with elements a and b we add the goal

(3) $\quad \leftarrow \text{Concat}(a.(b.e), c.e, z)$

Resolving (3) and (2) with $\{w/a, x/b.e, y/c.e, z/a.z_1\}$, where we have renamed the 'z' in (2) to be 'z_1' to make sure that the rule can be properly applied, we get

(4) $\quad \leftarrow \text{Concat}(b.e, c.e, z_1)$

Resolving (4) and (2) with $\{w/b, x/e, y/c.e, z_1/b.z_2\}$, where again the '$z$' in (2) is renamed, this time to be 'z_2', we get

(5) $\quad \leftarrow \text{Concat}(e, c.e, z_2)$

Resolving (5) and (1) with $\{v/c.e, z_2/c.e\}$ we get

(6) $\quad \leftarrow$

Tracing back through the substitutions: z_2 is $c.e$, z_1 is $b.(c.e)$, z is $a.(b.(c.e))$ as we would expect.

Again, we can form a tableau proof for this. In fact, exactly the proof above can be produced by the tableau method when extended beyond the method that we talked about in Chapter 4, that is if we add the idea of unification to guide the choice of substitutions for universally quantified variables. We shall not actually present these extensions to the tableau method, but the tableau that the extended method produces can be seen as one where many unproductive instances of universally quantified sentences, as produced by the universal rule in the basic method that we are using, can be ignored. The tableau shown in Figure 7.2 has had such economies made in it.

We get this by the following use of rules: (4) comes from (2) by several instances of the universal rule where w is a, x is $b.e$, y is $c.e$ and z is $b.(c.e)$; then (5) and (6) come from (4) by the $\rightarrow$-rule; then (7) comes from (3) by the universal rule using $a.(b.(c.e))$ for z, which causes a closure because (6) and (7) are contradictory; then (8) comes from (2) by the universal rule using e for x, $c.e$ for y, $c.e$ for z and b for w; then (9) and (10) come from (8) by another use of the $\rightarrow$-rule, and (10) contradicts (5) so we have another closure; finally (11) comes from (1) with $c.e$ for v in a use of the universal rule, and this closes the whole tableau as a result of a contradiction with (9).

We can see that the substitutions that are made, in particular the

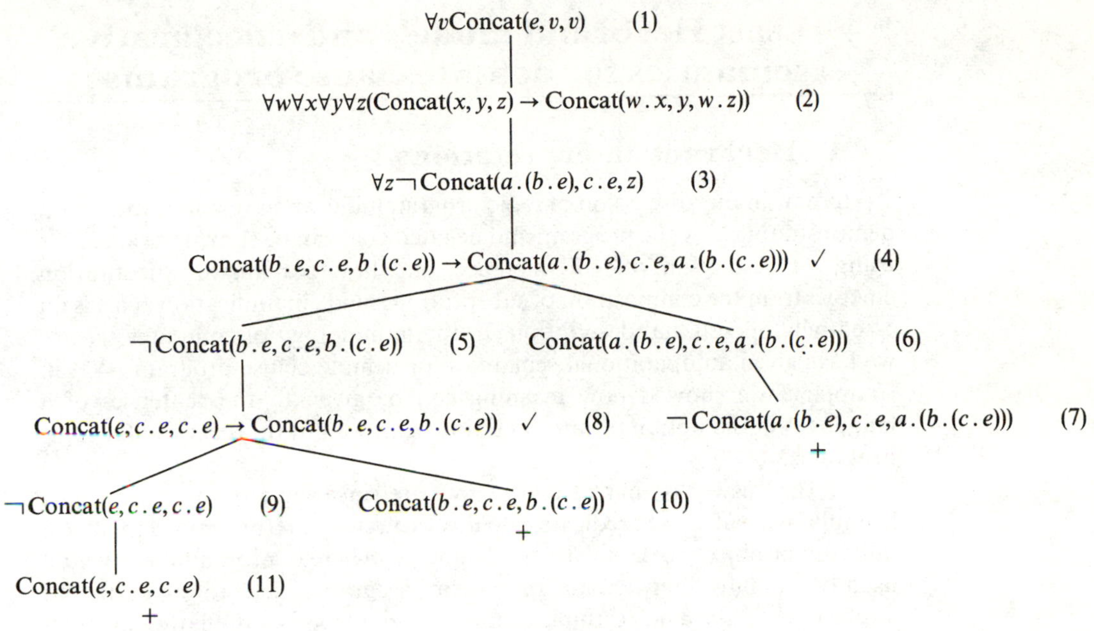

Figure 7.2.

substitution of $a.(b.(c.e))$ for the z in (3), are exactly those in the resolution proof. Again, this is as we would hope, since logically the two methods should give the same results given that they are both sound and complete.

Exercise 7.3

(a) Given the definite clause program

Concat$(e, v, v) \leftarrow$
Concat$(w.x, y, w.z) \leftarrow$ Concat(x, y, z)
Reverse$(e, e) \leftarrow$
Reverse$(w.x, z) \leftarrow$ Reverse(x, y), Concat$(y, w.e, z)$

use resolution to show that the following are logical consequences of the program. Calculate the answer substitution where relevant.

(i) Reverse$(a.(b.e), b.(a.e))$
(ii) Reverse$(a.(b.e), z)$

(b) Give corresponding tableau proofs for each of the above resolution proofs. As before, you should use the substitutions that the resolution proofs generate to guide your tableau proofs so as to make them as small as possible.

7.4 Least Herbrand models and a declarative semantics for definite clause programs

7.4.1 Herbrand interpretations

Up to now in our discussion of logic programming we have concentrated on demonstrating how the program and negated goal can be shown unsatisfiable using the resolution rule of inference, and how the answer substitution emerges from the composition of substitutions made in unification. This is an essentially operational description. In the terminology of computer science we have given an operational semantics for definite clause programs. Yet in Chapter 3 we showed how meaning can be given to the sentences of a language via the notion of interpretation. Can we not do the same for logic programs?

The answer is that of course we can, since logic programs are just formulas in a subset of predicate calculus. However, interpretations involve a universe of objects external to the language, whereas programmers would usually be unwilling to imagine their programs tied to a particular application. They tend to think of their programs as manipulating abstract objects such as numbers and sequences and being applicable to any problem whose data fits these abstractions. The analogy for logic programming would lead us to describe the meaning of such a program as relations over the terms that are involved in it. This raises an apparent problem since in Chapter 3 we were very insistent that the elements of the formal language and the elements of the domain were to be thought of as distinct.

We can bring language and interpretation closer together via the notion of **Herbrand interpretations**. Given a logic program P we can identify the set of all ground terms (terms not containing occurrences of variable symbols) that can be formed from the names and function symbols of P. For the language with constant symbols a and b, binary function symbol f and unary function symbol g, for example, this set is

$$\{a, b, f(a,a), f(a,b), f(b,a), f(b,b), g(a), g(b), f(g(a),b), f(g(g(b)), g(a)), \ldots\}$$

Clearly the set is enumerably infinite if there are any function symbols or infinitely many names.

We now construct an interpretation whose universe has an element corresponding to each element of this set. We may as well give them similar names, for example $\{\underline{a}, \underline{b}, \underline{f(a,a)}, \ldots\}$. This set is called the **Herbrand universe** for the program P. In practice, if we are careful to avoid confusion with terms of the language, we can omit the typographic distinction embodied in our underlining and just say that the Herbrand universe consists of the ground terms themselves.

We now define the **Herbrand base** for P as the set of all ground atoms, that is atomic formulas all of whose arguments are ground, that can be formed from the predicate symbols occurring in P and the Herbrand universe for P.

A Herbrand interpretation for P is then a subset of the Herbrand base, the elements of the subset being the ground atoms that are assumed to be satisfied in the interpretation.

A **Herbrand model** for P is a Herbrand interpretation that makes all the clauses of P true. For example, if P is the program

$$p(a)\leftarrow$$
$$q(b)\leftarrow$$
$$r(c)\leftarrow$$
$$p(x) \leftarrow q(x)$$
$$r(y) \leftarrow p(y)$$

then the Herbrand universe is $\{a, b, c\}$, and the Herbrand base is

$$\{p(a), p(b), p(c), q(a), q(b), q(c), r(a), r(b), r(c)\}$$

A Herbrand model for P is $\{p(a), p(b), q(a), q(b), r(a), r(b), r(c)\}$. The interpretation $\{p(a), p(b), q(a), q(b), r(a), r(c)\}$ would not be a Herbrand model for P (why not?).

Now what meaning can we attach to P? What can we say that the program P denotes? One possible answer is that the meaning of P is given by a Herbrand model for P. But which one? Some models are larger than strictly necessary to satisfy the program, as in the example above. The intersection of all Herbrand models for P is clearly also a Herbrand model for P and no Herbrand model can be smaller. We call this the **least Herbrand model** for P, and denote it by M_P. It can be shown that M_P is precisely the set of ground atoms that are the logical consequences of P, the set of ground goals that succeed by resolution with the clauses of P.

In complicated cases the least Herbrand model will not be obvious from inspection, but there is a recursive method of calculating it. Let the function f_P, which maps Herbrand interpretations into Herbrand interpretations, be defined as follows:

$$f_P(H) = H \cup \{B \mid B \leftarrow B_1, \ldots, B_n \text{ is a ground instance of a clause in } P \text{ and } \{B_1, \ldots, B_n\} \subseteq H\}$$

Then the least Herbrand model is the least fixed point of f_P, that is the smallest solution of the equation $f_P(H) = H$. For finite models this can be calculated by setting $H_{n+1} = f_P(H_n)$, where $H_0 = \{\ \}$, and finding the smallest n for which $H_{n+1} = H_n$. For the program P above, we have

$$H_1 = f_P(\{\ \}) = \{p(a), q(b), r(c)\}$$
$$H_2 = f_P(H_1) = \{p(a), q(b), r(c), p(b), r(a)\}$$
$$H_3 = f_P(H_2) = \{p(a), q(b), r(c), p(b), r(a), r(b)\}$$
$$H_4 = f_P(H_3) = \{p(a), q(b), r(c), p(b), r(a), r(b)\}$$

So the least Herbrand model for P is given by H_4 and it is obvious that this is the complete set of goals that would succeed, given P. It seems reasonable to view the least Herbrand model of a logic program as canonical, in this sense, and to say that that is what the meaning of the logic program is.

7.4.2 General clauses and theorem proving

Herbrand interpretations have a significance that extends beyond definite clause programs. One can define a **clause** more generally as a universally quantified finite disjunction of literals. **Literals** are atomic formulas or their negations. The clauses that can appear in definite clause programs, or their goals, are special cases. Using the equivalence

$$(Q_1 \wedge Q_2 \wedge \ldots \wedge Q_n \rightarrow P) \leftrightarrow (\neg Q_1 \vee \neg Q_2 \vee \ldots \vee \neg Q_n \vee P)$$

it can be seen that definite clauses have precisely one unnegated literal and goals have none. (Clauses with at most one unnegated literal are called **Horn clauses.**)

It can be shown that any sentence $\mathcal{S}$ of classical predicate calculus can be transformed into a set of clauses $C_{\mathcal{S}}$ such that if $\mathcal{S}$ is satisfiable then $C_{\mathcal{S}}$ has a Herbrand model. This means that if $C_{\mathcal{S}}$ can be shown by resolution to have no Herbrand model then $\mathcal{S}$ is unsatisfiable. Resolution therefore has application wider than logic programming, in that, like the tableau method, it can be used to test the validity of any sentence in predicate calculus. Robinson's (1979) book is the definitive reference. For a full description of definite clause programming and its theory, see Lloyd (1984).

Exercise 7.4 Show that the sentence

$$P(a) \wedge \exists x(\neg P(x))$$

has a model, but does not have a Herbrand model.

SUMMARY

- The idea behind logic programming is that the program is a specification, in a logical language, of what constitutes a solution to the problem, and the computational mechanism (operational semantics) is then provided by a system of deduction for that language.

- The logic programs in this chapter are *definite clause programs*. A definite clause is a universally quantified disjunction of one or more literals, only one of which is unnegated. Definite clauses can be compared with procedure definitions in a procedural language. Definite clause programs are executed by adding a *goal*, a clause in which there are no unnegated literals.
- The system of deduction in this form of logic programming is *resolution*. Like the tableau method, resolution works by refutation. Unlike the tableau method, it requires the set of input sentences to be in a special form – a universally quantified conjunction of disjunctions – known as *clause form*. The resolution rule of inference sanctions the derivation, from a set of clauses, of another set having the property that, if the original set is satisfiable, so is the derived set. If falsity in the shape of the empty clause can be derived, the original set is unsatisfiable and the entailment from which they were constructed is therefore valid.
- Resolution in the predicate calculus involves *unification*. Unification is the process of finding substitutions that make two terms containing free variables identical. A *substitution* is a function from variable symbols to terms and can be represented as a set of variable bindings. Algorithms have been given for *composition* of substitutions and for finding the *most general unifier* (mgu) of two terms. The mgu is unique up to permutation of variable names.
- *Herbrand interpretations* can be used to give a semantics for logic programs. A Herbrand interpretation is based on the set of all ground atoms that can be constructed from the names, function symbols and predicate symbols in the program. The *least Herbrand model* of a program can be calculated as the least fixed point of a certain function that maps Herbrand interpretations into Herbrand interpretations.
- Any predicate calculus sentence can be transformed into a set of clauses such that if the original sentence is satisfiable, the corresponding clauses have a Herbrand model. In this case, for some sentences, resolution may not terminate. If the clauses can be shown, by resolution, to have no Herbrand model then the original sentence is unsatisfiable. So, like the tableau method, resolution can be used to test, by refutation, the validity of any entailment in predicate calculus.

Non-standard Logics

8.1 Introduction

In this chapter we shall look at other systems which formalize reasoning. We pre-empted this discussion a little when we introduced intuitionistic logic in Chapter 5, and we shall return to this later. The other systems that we present here will be different from the logics, apart from intuitionistic logic, that we have seen so far, which we call classical logics, in that other sorts of valid reasoning will be treated. By other sorts of reasoning we mean both extensions to the sorts of arguments that we have looked at before and arguments based on different fundamental assumptions.

8.2 Necessity and possibility

So far, we have only formalized reasoning in which relations have a truth-value timelessly. That is, when we say 'x is red' we do not refer to when it is red or for how long it is red or how I know that it is red. However, we often need to reason in situations where states of affairs change over time (especially in computing). Two important time- or state-dependent qualities of relations tell us whether a relation is necessarily true, that is true at all times or in all situations, or whether a relation is possibly true, that is there is at least one moment or state when or where the relation holds.

Whether a relation is necessary or possible is itself a proposition on the predicate, that is has a truth-value, which expresses something about the mood or manner or mode of the predicate. This leads to the name **modal logic** for a system of reasoning in which the ideas of necessity and possibility occur.

8.2.1 Possible worlds

To begin to formalize such a logic we need to extend our semantics. Consider the sentence 'the sky is blue'. Clearly we have (by common experience) that 'the sky is necessarily blue' is false (sunsets?) and 'the sky is possibly blue' is true. So, there are situations or possible worlds in which 'the sky is blue' is true, and some in which it is false.

By **possible worlds** we mean not only actually occurring worlds but any logically possible situation, that is one in which no contradictions arise. So, any imagined world counts as a possible world, as long as it is not logically impossible, that is contradictory. Thus, possibility is not the same as conceivability, it is a more general notion, and not all possible worlds are actual either. For instance, 'London buses are black' is true in some imaginable world. The possibility is not ruled out just because it is not actually the case. In the future of our actual world (in which London buses are red) someone might decide to re-paint them. However, a proposition such as 'the sky is blue and the sky is not blue' cannot be true in any world since it is a contradiction.

For example consider the sentence '$0 = 1$', where we assume that all the terms in the sentence have their usual meanings. We ask whether there is a possible world in which this sentence is true. To see how to decide this question imagine the proposition 'there are two objects on the table' in the case where there are two objects on the table at which you might be sitting. Then, if we also accept the truth of the sentence '$0 = 1$' and since $2 + 0 = 2 + 0$ it follows that $2 + 0 = 2 + 1$ and hence $2 = 3$. So, the proposition 'there are not two objects on the table' is true in exactly the same circumstances as the proposition 'there are two objects on the table'. Clearly, these are contradictory.

This contradiction came about by using true facts of arithmetic together with the sentence '$0 = 1$'. The only way out of this contradiction, then, is to deny that this sentence is true in the world and, by a similar argument, in any possible world, since no possible world can be contradictory and the inclusion of the sentence '$0 = 1$' as true in any world gives rise to contradiction.

Exercise 8.1 In the following exercise, all the terms that appear should be understood with their usual meanings. For each sentence, say whether or not there is a possible world in which it can be true:

(a) Mount Snowdon is higher than Mount Everest.

(b) $2 + 2 \neq 4$.

(c) A yard is shorter than a metre, a metre is shorter than a chain and a chain is shorter than a yard.

(d) I believe that there is a greatest natural number.

(e) I know that there is a greatest natural number.

8.2.2 Contingency

A sentence or proposition which can be either true or false, depending on the situation it is evaluated in, is called **contingent**. Any proposition which has the same truth-value no matter what situation it is evaluated in is called **non-contingent**. For example, 'the sky is blue' is contingent since, for example, on Mars it is pink, that is the sentence is false, whereas on Earth the sentence can be true. On the other hand 'all yellow things are yellow' is true no matter what situation or possible world it is in, that is it is non-contingent.

Exercise 8.2 In each of the following, say whether the sentence is contingent or non-contingent. If you think that it is non-contingent, say whether it is true or false.

(a) All uncles are males.

(b) All males are uncles.

(c) All black birds are black.

(d) All blackbirds are black.

(e) No toadstools are poisonous.

(f) All coloured things are red.

8.3 Possible world semantics

This idea of possible worlds or situations has been taken up and used as one basis for a formalization of modal logic. We shall use this basis too. However, unlike classical logic, formalizations and proof systems for modal logics (among others) are still a matter of research and our presentation is by no means the only one or universally agreed to be the correct one (whatever that means).

We use an extension of the language of classical (propositional) logic, so we have $\wedge$, $\vee$ and so on.

If we know that A is necessarily true then we are saying that whatever worlds we can 'get to' or 'imagine' in some way – we shall say **access** – then A is true in each of them. So, since we know that 'anything red is necessarily red' we are saying that in any world that we can access from our world 'anything red is red' is true.

Similarly, since 'possibly the sky is blue' is true we are saying that there is at least one world that we can access from our world in which 'the sky is blue' is true.

8.4 Frames, interpretations and models

We wish to formalize the ideas above. We have a set of possible worlds P and a relation, the accessibility relation R, which tells us which worlds are accessible from which others. (You may also, though rarely, see the term **reachability**, instead of accessibility). So, if W and V are possible worlds, that is $W, V \in P$, then $R(W, V)$ means that V is accessible from W. The exact definition of any particular R will determine what 'sort' of accessibility we mean, as we shall see later. Thus, non-accessible worlds are in some way irrelevant to us in the current world.

Definition 8.1: If P is a set of possible worlds and R is an accessibility relation then $\langle P, R \rangle$ is a **frame (of reference)**.

Definition 8.2: Let $\langle P, R \rangle$ be a frame. An **interpretation** in $\langle P, R \rangle$ is a function v such that

$$v: P \times L \rightarrow \{\text{true}, \text{false}\}$$

where L is a set of propositional letters.

So, an interpretation tells us whether a given proposition (from L) is true in a possible world (from P). Note how this extends our previous definition because now we need to take into account the world in which the sentence is being considered as well the sentence itself.

For example: let $P = \{\text{today}, \text{tomorrow}\}$ and $L = \{P_1, P_2\}$. Then let
$v(\text{today}, P_1) = \text{true}$
$v(\text{tomorrow}, P_1) = \text{false}$
$v(\text{today}, P_2) = \text{false}$
$v(\text{tomorrow}, P_2) = \text{true}$

and let the accessibility relation be {(today, tomorrow), (today, today), (tomorrow, tomorrow)}; then, informally, you might now see that, as far as 'today' is concerned, 'possibly P_1' and 'possibly P_2' are true while 'necessarily P_1' and 'necessarily P_2' are false. What we have to do is to formalize this notion, which we now do.

8.4.1 Extending the language

We extend our alphabet and language with two new symbols. $\Diamond$ (diamond) means 'possibly' and $\Box$ (box) means 'necessarily'. Then, if $\mathscr{S}$ is any sentence, so are $\Diamond\mathscr{S}$ and $\Box\mathscr{S}$. So, in our example we have $\Diamond P_1$, $\Diamond P_2$, $\neg\Box P_1$ and $\neg\Box P_2$.

Definition 8.3: For a frame $\langle P, R\rangle$, we write $W \Vdash_v \mathscr{S}$ iff $\mathscr{S}$ is true in the interpretation v at world $W \in P$, that is $W \Vdash_v \mathscr{S}$ iff $v(W, \mathscr{S}) = \text{true}$ for some v. When v is clear from the context that we are in we drop it. We read '$W \Vdash \mathscr{S}$' as 'W forces $\mathscr{S}$'.

Definition 8.4: If v is an interpretation in the frame $\langle P, R\rangle$ of a set of sentences S at any world $W \in P$ then, given the above definition, we have the following for any $\mathscr{S}, \mathscr{T} \in S$:

(m_1) $W \Vdash (\mathscr{S} \wedge \mathscr{T})$ iff $W \Vdash \mathscr{S}$ and $W \Vdash \mathscr{T}$

(m_2) $W \Vdash (\mathscr{S} \vee \mathscr{T})$ iff $W \Vdash \mathscr{S}$ or $W \Vdash \mathscr{T}$, or both

(m_3) $W \Vdash (\mathscr{S} \rightarrow \mathscr{T})$ iff not $W \Vdash \mathscr{S}$ or $W \Vdash \mathscr{T}$, or both

(m_4) $W \Vdash \neg\mathscr{S}$ iff not $W \Vdash \mathscr{S}$

(m_5) $W \Vdash \Box\mathscr{S}$ iff for all $V \in P$ such that $R(W, V)$ we have $V \Vdash \mathscr{S}$

(m_6) $W \Vdash \Diamond\mathscr{S}$ iff there is a $V \in P$ such that $R(W, V)$ and $V \Vdash \mathscr{S}$

EXAMPLE 8.1

Going back to our previous example, we can illustrate these definitions. We will take P and L as before and re-define R to be the equivalence relation

{(today, tomorrow), (tomorrow, today), (today, today), (tomorrow, tomorrow)}

We have

today $\Vdash P_1$
tomorrow $\Vdash P_2$
today $\Vdash \neg P_2$
tomorrow $\Vdash \neg P_1$

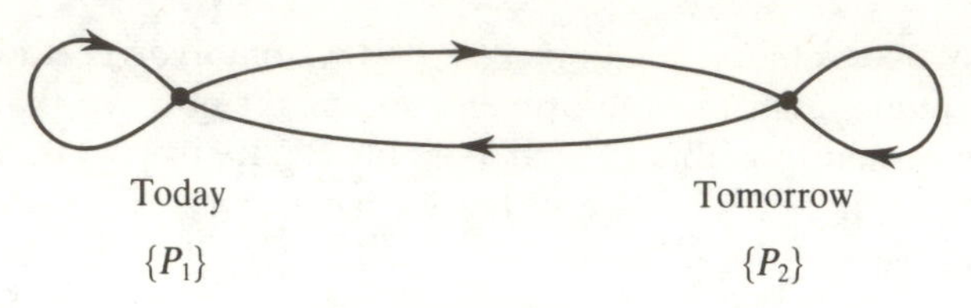

Figure 8.1.

Since it is not the case that for all $U \in P$ such that $R(W, U)$ we have $U \Vdash P_1$ (because tomorrow $\in P$, R(today, tomorrow) and not tomorrow $\Vdash P_1$) it follows that not today $\Vdash \Box P_1$. Similarly, we can show that

not today $\Vdash \Box P_2$
not tomorrow $\Vdash \Box P_1$
not tomorrow $\Vdash \Box P_2$

However, since today $\in P$, R(today, today) and today $\Vdash P_1$ we have today $\Vdash \Diamond P_1$. Also, since today $\in P$, R(tomorrow, today) and today $\Vdash P_1$, we have tomorrow $\Vdash \Diamond P_1$. Similarly

today $\Vdash \Diamond P_2$
tomorrow $\Vdash \Diamond P_2$

We can illustrate these facts by using a diagram, a **possible worlds diagram**, which is essentially a labelled, directed graph. The nodes are intended to be possible worlds, so they are labelled with their names. Also, at each node we write the propositions which are true there. The arcs are arrows which show which worlds (nodes) are directly accessible from which others. Therefore, the above example will be as shown in Figure 8.1. There we see that the accessibility relation is reflexive, transitive and symmetric.

EXAMPLE 8.2

As another example consider the frame $\langle P, R \rangle$ given by

$$P = \{W_1, W_2\}$$
$$R = \{(W_1, W_2), (W_1, W_1), (W_2, W_2)\}$$

Consider the sentence $A \rightarrow \Box \Diamond A$ in the interpretation v given by

$$v(W_1, A) = \text{true}$$
$$v(W_2, A) = \text{false}$$

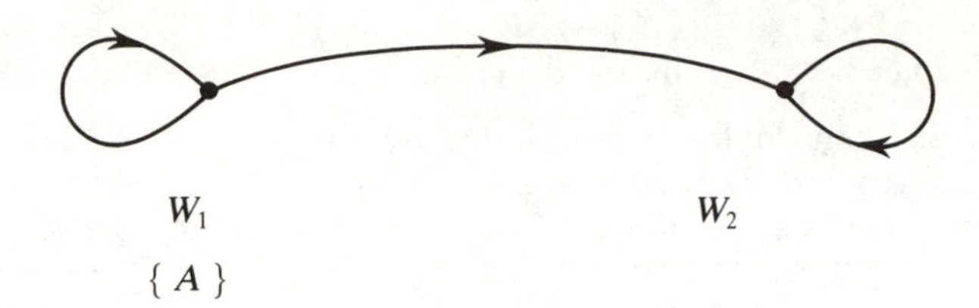

Figure 8.2.

Then, $W_1 \Vdash A$. Now, what about $W_1 \Vdash \Box\Diamond A$? Is it true that for all $U \in P$ such that $R(W_1, U)$ that $U \Vdash \Diamond A$, that is, is it true that $W_1 \Vdash \Diamond A$ and $W_2 \Vdash \Diamond A$, that is, is it true that

(1) there is a $U \in P$ such that $R(W_1, U)$ and $U \Vdash A$ and

(2) there is a $U \in P$ such that $R(W_2, U)$ and $U \Vdash A$?

Well, (1) has the answer yes since $W_1 \in P, R(W_1, W_1)$ and $W_1 \Vdash A$. However, (2) has the answer no. This is because although $W_1 \in P$ and $W_1 \Vdash A$, it is not true that $R(W_2, W_1)$. And, although $W_2 \in P$ and $R(W_2, W_2)$, it is not true that $W_2 \Vdash A$.

So, we finally have that not $W_1 \Vdash \Box\Diamond A$, so not $W_1 \Vdash A \rightarrow \Box\Diamond A$. Figure 8.2 shows the possible worlds diagram that describes this.

Definition 8.5: Let $\mathsf{F} = \langle P, R\rangle$ be a frame, with P a set of possible worlds and R an accessibility relation, and let v be an interpretation in F. Also, let $\mathscr{S}$ be a sentence. Then v is a **model of $\mathscr{S}$ in F** iff $v(W, \mathscr{S}) =$ true for all $W \in P$. $\mathscr{S}$ is **satisfied** by v iff $v(W, \mathscr{S}) =$ true for some $W \in P$. So, $\mathscr{S}$ is **satisfiable** iff there is an interpretation in F such that $v(W, \mathscr{S}) =$ true for some $W \in P$. $\mathscr{S}$ is **true in v in F** iff v in F is a model of $\mathscr{S}$. $\mathscr{S}$ is **valid in F** iff all interpretations in F are models of $\mathscr{S}$, that is $\mathscr{S}$ is true in all interpretations in F.

Going back to the second example above, $A \rightarrow \Box\Diamond A$ is not valid in $\langle P, R\rangle$ since we have given an interpretation in $\langle P, R\rangle$ which is not a model, that is we have an interpretation in which the sentence is not true for all possible worlds in P.

Exercise 8.3

(a) Draw the possible worlds diagrams, as exemplified above, for the following frames:

(i) $Pa = \{W_1, W_2, W_3\}, Ra = \{(W_1, W_1), (W_2, W_2), (W_3, W_3)\}$

(ii) $Pb = \{W_1, W_2, W_3\}, Rb = \{(W_1, W_2), (W_2, W_1), (W_3, W_3)\}$

(iii) $Pc = \{W_1, W_2\}, Rc = \{(W_1, W_2), (W_2, W_1), (W_1, W_1), (W_2, W_2)\}$
(iv) Pd = set of natural numbers, Rd = the less than relation over Pd

(b) Draw the possible words diagram for frames given by the following:
(i) $P = \{W_1, W_2\}$, R is an equivalence relation over P;
(ii) with P and R as in (i), but a different frame.

8.5 Truth functionality and modal logic

When we say that a logic is **truth-functional** we mean, essentially and as you should recall, that its semantics can be given by truth-tables. Putting this another way, it implies that the meaning of a sentence can be calculated from the meanings of its subsentences. However, modal logics do not have this property. Thus, they are non-truth-functional, as we shall see. This means that we cannot use the simple ideas of truth-tables to give the meanings of such logics, so the (relatively) complicated methods introduced above must be used.

For example, the operator $\diamond$ is not truth functional (unlike $\neg$ which is) because the truth-value of $\diamond A$ does not depend only on the truth-value of A. If A is true, then clearly $\diamond A$ is true too. However, if A is false the answer to the question must be 'it depends ...'. When pressed further we might say 'it depends on the situation'. Exactly! The truth-value of $\diamond A$ depends not only on the truth-value of A but also on the situation, that is on the frame in which the question is asked.

If we try to fill in its truth-table we get

$\diamond$	
t	t
f	?

where '?' stands for 'it depends ...'.

Another way of seeing that there is no truth-table for $\diamond$ is to consider the fact that $\diamond$ is a unary operator and to write down all the truth-tables, since there are only four. However, two of them are 'used' in the sense that one of them is the truth-table for $\neg$ and another is just the identity function (which $\diamond$ and $\square$ certainly are not). The other two give 't' for all possibilities and 'f' for all possibilities. Clearly, these will not do either. So, there are not enough truth-tables to give one each to $\diamond$ and $\square$.

A further way of exploring this problem is to introduce more than two truth-values, for example true, false and undefined. This idea was worked on extensively by many people. A major example is the three-valued logic of Lukasiewicz.

8.6 Systems of modal logic

Obviously, as the frame varies, even for a fixed language, the behaviour of the logic will change, that is we shall have different logics. It turns out that certain simple ways of axiomatizing the logics, which we have not looked at for modal logics, correspond to certain simple and familiar properties of the accessibility relation. A series of such axiomatizations were studied in the past by Lewis (Lewis and Langford, 1932) and the elements of the series were named S1, S2, ... that is system 1, system 2, ... They were generated by taking some initial set of axioms and adding new axioms to get new logics with particular properties. We shall be looking at S4 in the next section. In S4 the accessibility relation is reflexive and transitive. The system after S4, that is S5, is such that its accessibility relation is an equivalence relation, that is it is reflexive, transitive and symmetric. Though we shall say no more about these systems other than S4, it is remarkable that they can be characterized in such simple terms.

8.7 A tableau system for S4

8.7.1 Introduction

We have chosen to use the possible worlds idea (originally due to Kripke (1963)) to present a semantics for modal logics. One component of such a semantics is the accessibility relation. The choice of this relation, and in particular its properties, reflects just what sort of reasoning we are doing. For instance, if the relation is not symmetric then we cannot be reasoning about a system in which 'going back to where we have been' is possible, that is if I pass from world W to world V then I cannot pass from V back to W, in general. From now on we shall concentrate on a logic whose accessibility relation is reflexive and transitive. This is the logic S4 as mentioned in the previous section. Our goal here is to develop a tableau system which, as before, will give us a mechanical way to test the validity (in S4) of sentences in our language. We shall use the possible worlds interpretation to do this.

First, remember that the basic idea behind tableaux is to explore all possible ways in which a sentence or set of sentences can be satisfiable. If we ask the right question, we can thus devise a system which allows us to decide whether a sentence is valid. By valid here we mean valid in any S4 logic, that is we specialize the definition above to

> **Definition 8.6:** A sentence $\mathscr{S}$ is **S4-valid** iff $\mathscr{S}$ is true in all interpretations in any frame $\langle P, R \rangle$ where R is reflexive and transitive. We write this as

$$\models_{S4} \mathscr{S}$$

We shall say just 'valid' in the future and write just

$$\models \mathscr{S}$$

Thus, above we showed that it is not the case that

$$\models (A \to \Box\Diamond A)$$

that is $A \to \Box\Diamond A$ is not S4-valid.

This definition also extends to sets of sentences in the usual way.

8.7.2 Testing validity

Now, we want to devise a way to answer questions such as 'is it the case that $\models \mathscr{S}$?' As before in the classical case, what we do is to try to find all possible ways in which $\neg\mathscr{S}$ is (S4-) satisfiable. If no such way exists then we conclude that $\models \mathscr{S}$. It turns out that the rules for non-modal operators (that is those other than $\Box$ and $\Diamond$) are much as before. We need new rules to deal with modal sentences. (We are essentially presenting Kripke's original work as extensively developed by Fitting (1983)). Consider the case for a sentence of the form $\Box\mathscr{S}$. How can we express (in our tableau style) all possible ways in which this can be true? Well, it says that $\mathscr{S}$ is necessarily true, that is that $\mathscr{S}$ is true in all worlds accessible from our current one. So, if we were to move to a representative accessible world, $\mathscr{S}$ would be true there.

Now, what about $\Diamond\mathscr{S}$? We can express the truth of this in all possible ways by thinking of moving to an arbitrary accessible world in which $\mathscr{S}$ is true and which includes just what must be there. Putting this another way, in moving to this other world we need to delete all information that does not have to be there, that is all non-necessary information. So, in moving between worlds we should delete all sentences not of the form $\Box\mathscr{T}$ or $\neg\Diamond\mathscr{T}$, since no sentence other than these has to be true in every accessible world.

The reason why this is correct is that, if our sentence $\neg\mathscr{S}$ is not satisfiable in this 'minimal' accessible world, then it is certainly not satisfiable in any world which has more things true in it. So if, when we need to change worlds, we always change to one which only has in it the things which must be true, and our test sentence is unsatisfiable in an arbitrary such world, then it must be unsatisfiable in any accessible worlds (since these will have either the same things true in them or more things true in them, but never less, by definition).

We shall give the extra rules needed and then give some more justification. Written in the pictorial way, as previously, we have four more rules:

Definition 8.7: The **non-deleting modal rules** can be pictured as (where $\mathscr{S}$ is any sentence)

$$\begin{array}{cc} \Box\mathscr{S} & \neg\Diamond\mathscr{S} \\ \mathscr{S} & \neg\mathscr{S} \end{array}$$

and the **deleting modal rules** can be pictured as

$$\begin{array}{cc} \Diamond\mathscr{S} & \neg\Box\mathscr{S} \\ \mathscr{S} & \neg\mathscr{S} \end{array}$$

but in these two cases delete from the paths that have just been extended all sentences except those of the form $\Box\mathscr{R}$ and $\neg\Diamond\mathscr{R}$.

EXAMPLE 8.3

Consider

$$\models A \rightarrow \Box\Diamond A$$

the tableau for which is in Figure 8.3. As usual, we start with the initial sentence which is the negation of the one whose validity we are testing, that is we start with the initial sentence $\neg(A \rightarrow \Box\Diamond A)$. Here, (2) and (3) came from (1) by the usual rule for $\neg\text{–}\rightarrow$. Then, (4) came from (3) by the rule for $\neg\text{–}\Box$, which also caused the crossing out of sentences (1), (2) and (3), since none of them is of the form which we have to save, that is none of them has to be true in the 'minimal' accessible world. Finally (5) came from (4) by the rule for $\neg\text{–}\Diamond$.

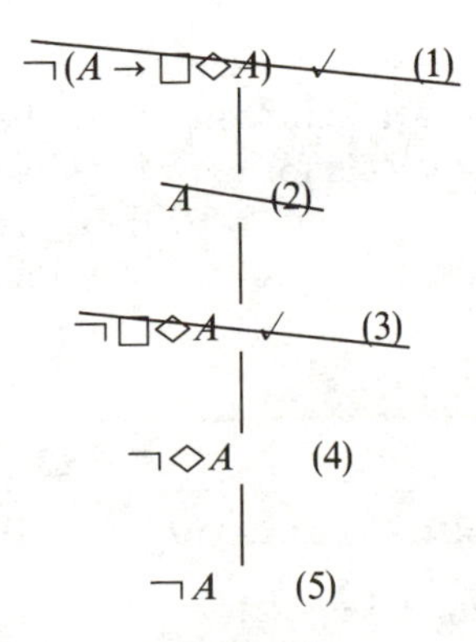

Figure 8.3.

The tableau does not close here because the sentences true in the current world, that is those not crossed out, are not contradictory. Since the tableau is complete but not closed, we say that, having elaborated the tableau in all possible true ways, there is one which is not contradictory, namely the one in which there is a world in which $\neg A$ is true. If you look at the way we dealt with this example before you will see that this is exactly the frame we used.

EXAMPLE 8.4

As another example consider

$$\models(\Box\neg A \leftrightarrow \neg\Diamond A)$$

Informally, we can see that this says that if A is necessarily false then it is not possible that it is true, and vice versa. The tableau method gives the tableau shown in Figure 8.4.

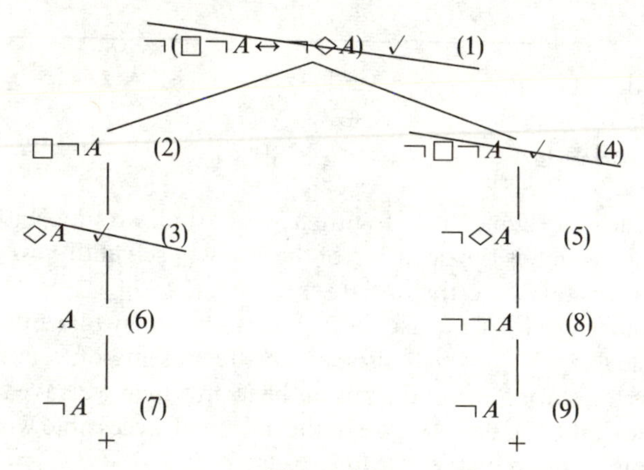

Figure 8.4.

Here (1) is as usual the negation of the sentence whose validity is being tested. (2), (3), (4) and (5) come from the rule for $\neg$–$\leftrightarrow$. Then, (6) comes from (3) by the rule for $\Diamond$, which also causes (1) and (3) to be deleted. Then (7) comes from (2) by the $\Box$ rule. This path then closes because it contains both A and $\neg A$. (8) comes from (4) by $\neg$–$\Box$ and (9) comes from (5) by $\neg$–$\Diamond$. This path then closes too since $\neg A$ and $\neg\neg A$ are contradictory.

8.7.4 Cycles and backtracking

With these new rules for modal operators, we need to take care about when we use them. In fact, they must only be used after all possible sentences with

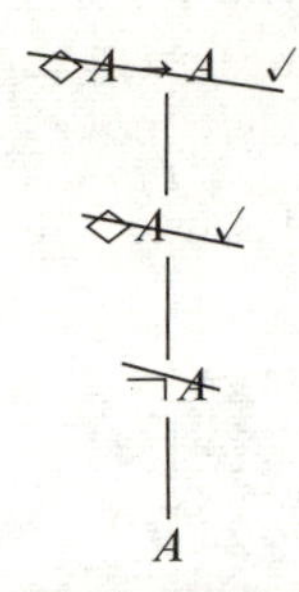

Figure 8.5.

non-modal operators (that is the connectives) at the topmost level have been used. The following example shows that we get wrong results otherwise. We know that

$$\{\Diamond A \to A, \Diamond A\} \models A$$

since this is just an instance of $\{\mathscr{S} \to \mathscr{T}, \mathscr{S}\} \models \mathscr{T}$, but if we do not take note of the point above we might get the tableau shown in Figure 8.5, which does not close.

A further complication arises when two (or more) sentences of the form $\Diamond\mathscr{S}$ occur on the same path. Consider the tableau for the valid sentence

$$\neg\Box\neg(\Box A \lor \Box\neg\Box A)$$

as given in Figure 8.6.

In constructing this tableau we chose sentences (1) and (1′) to be elaborated at each choice point, that is when we first had to choose between (1) and (2) and then between (1′) and (2′). When we reach the state shown in the figure we can stop building the tableau because it is clear that we have reached a cycle, that is if we keep making the same choices then the path will never be finished and so never closes. Hence, we would conclude that the initial sentence is satisfiable, that is that $\neg\Box\neg(\Box A \lor \Box\neg\Box A)$ is invalid. But we know that this is not the case, so we need to do something to stop this wrong conclusion being reached.

The problem is that we have not explored all possible ways in which the initial sentence, and its subsentences, can be elaborated. For instance, we have at no point chosen sentence (2) as the one to be used. One way of making sure that all possibilities are eventually tried is to remember where a choice is made and to use backtracking.

So, if we detect a cycle then we backtrack to the previous choice point, make a different choice, if possible, and carry on. If no other choice was possible, that is all the others have been tried, then we backtrack to an even

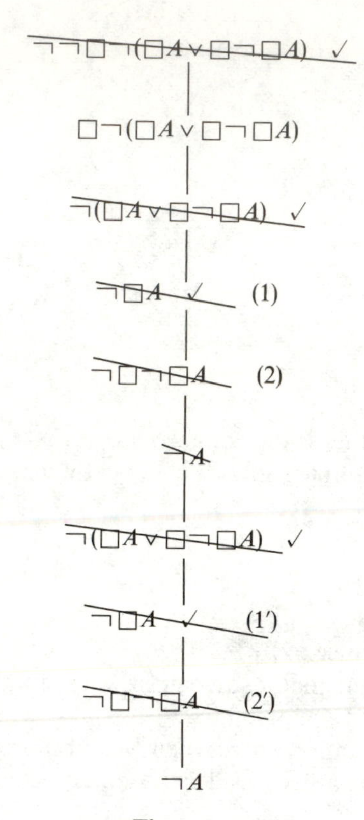

Figure 8.6.

earlier choice point and make another choice if possible. If all choices have been made and they lead to non-closing paths then, and only then, can we conclude that the initial sentence is invalid.

Going back to the example we would backtrack to the last choice point, that is where we chose (1′), and this time choose (2′). As you can see in Figure 8.7 the path does close now because the starred sentences are contradictory. As you can check for yourself, an even shorter proof could have been obtained by choosing sentence (2) at the first choice point.

Finally, if a path p has any sentence $\mathscr{S}$ deleted from it and part of p is common to any other path q from which $\mathscr{S}$ should not be deleted, add a new occurrence of $\mathscr{S}$ to q.

Having taken all these points into consideration we can write down an algorithm (Figure 8.8) for constructing tableaux as we did in the classical case. This will be an easy adaptation of the original one as you can see.

Note here that the non-deleting modal rules do not lead to the sentence being marked as used. This is so since if a sentence is of the non-deleting sort, that is either $\Box\mathscr{S}$ or $\neg\Diamond\mathscr{S}$, then either $\mathscr{S}$ or $\neg\mathscr{S}$ is true in all possible worlds (respectively), including any minimal ones we might move to, so they should be available for elaboration in all those worlds. This is rather similar to the case for the universal quantifier in classical logic.

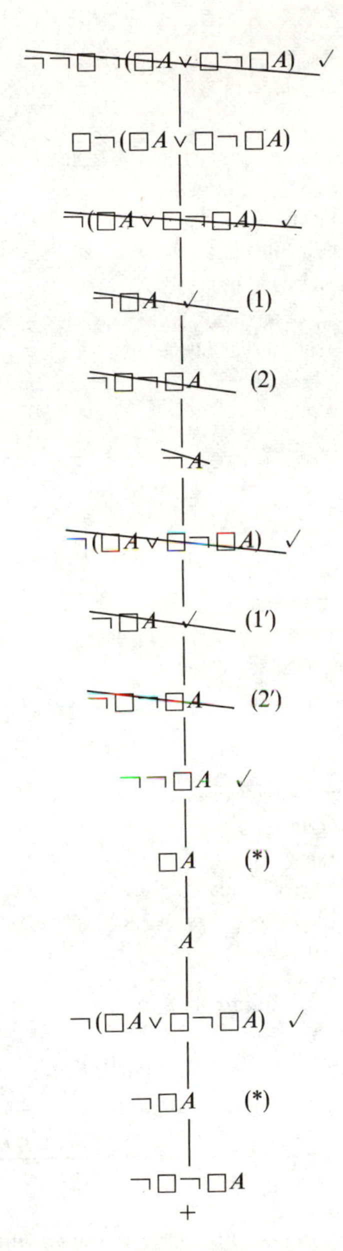

Figure 8.7.

```
begin
   repeat
      changes:= false
      close each path that contains a sentence and its negation
```

Figure 8.8. **(continues)**

```
if all paths are closed
then deliver "S4-valid"
else
  if a connective rule can be applied to a sentence
  then
    changes:= true
    apply the appropriate rule
    mark the sentence as used
  else
    if a non-deleting modal rule can be used
    then
      apply the rule
      changes:= true
    else
      if a deleting rule can be used
      then
        apply a deleting modal rule
        mark the sentence as used
        remember this as a choice point
        do the necessary deletions
        changes:= true
  if a cycle is detected
  then
    backtrack to last choice point
    changes:= true
until changes is false
  backtrack to last choice point (if any)
deliver "not S4-valid"
end
```

Figure 8.8 *(cont.)*

Once again we can show, though we will not here, that this algorithm is sound and complete.

Exercise 8.4

(a) Use the tableau method for S4 to decide the validity of the following sentences:

(i) $\Box A \rightarrow A$
(ii) $A \rightarrow \Box A$
(iii) $\Diamond(A \wedge B) \rightarrow (\Diamond A \wedge \Diamond B)$
(iv) $\Diamond A \rightarrow A$
(v) $\Box(A \vee \neg A)$
(vi) $\Box(A \wedge \neg A)$
(vii) $\Box(\Box A \vee \neg \Diamond A)$
(viii) $\Diamond\Diamond A$

(b) Check your answers to (i), (ii), (iv), (v) and (viii) above by using the definitions of the semantics of the modal operators as given in Section 8.4.

8.8 One use for modal logic in programming

In this section we shall briefly illustrate how modal logic has been used to get some grasp of the problem of expressing properties of programs (as you would want to if you are going to reason about their correctness). This example is one among many given by Zohar Manna and Amir Pnueli (Manna and Pnueli, 1981).

First, we need to set the scene. We shall consider sequential computation as described by a language which manipulates states. A state will be a pair $(\lambda, \mathbf{v})$ where λ is the current location and $\mathbf{v}$ is a vector of current identifier values, that is an environment. Each statement of the program will be linked with the next by a location. So, for instance, we can write

$$
\begin{array}{ll}
 & y_1 := 2 \\
\lambda_0: & \text{print}(y_1) \\
\lambda_1: & y_1 := y_1 + 1 \\
\lambda_2: & y_2 := 2 \\
\lambda_3: & \text{if}\,(y_2)^2 > y_1 \text{ then goto } \lambda_0 \\
\lambda_4: & \text{if}\,(y_1 \bmod y_2) = 0 \text{ then goto } \lambda_1 \\
\lambda_5: & y_2 := y_2 + 1 \\
\lambda_6: & \text{goto } \lambda_3
\end{array}
$$

which we call PR. The ys are identifiers, so their values are what are stored in $\mathbf{v}$ (in order, that is value of y_1, value of y_2 and so on), and the λs are locations. Therefore, an example of a state is $(\lambda_1, \langle 2, 2 \rangle)$.

Then, our set of possible worlds will be a set of such states and the accessibility relation holds between two states $S = (\lambda_i, \mathbf{v})$ and $T = (\mu_k, \mathbf{w})$ iff there is a statement, whose location is λ_i, which being executed in state S results in state T. So in our example $(\lambda_4, \langle 3, 2 \rangle)$ is related to $(\lambda_5, \langle 3, 2 \rangle)$ and this state is in turn related to $(\lambda_6, \langle 3, 3 \rangle)$ and so on.

We also define a useful predicate 'at λ' which is true iff in the current state the current location is λ.

Now, the program PR prints an infinite sequence of successive prime numbers 2 3 5 7 11 13 17 19.... If we have a predicate 'prime(x)' which is true iff x is a prime number then using our knowledge of modal logic we can make statements about the program. For instance, part of the statement that the program is correct is

$$\models \Box(\text{at } \lambda_0 \rightarrow \text{prime}(y_1)) \qquad \textbf{(8.1)}$$

that is it is necessarily the case that if we are in a state whose current location is λ_0 then the value of the first current identifier is a prime number. We might paraphrase this as 'nothing but primes are printed'. But there is much more to be said before we have stated that PR prints the required sequence. For instance, the sequence 3 17 2 5 19... also satisfies relation 8.1. The property that every prime number is printed could be expressed by

$$\models \forall u[(\text{at } \lambda_0 \wedge y_1 = 2 \wedge \text{prime}(u)) \rightarrow \Diamond(\text{at } \lambda_0 \wedge y_1 = u)] \tag{8.2}$$

that is if we are ever in the state where the current location is λ_0 and y_1 is 2 (which we are at the start of an execution of PR) and u is any prime number then there is at least one state in which we are at λ_0 and y_1 is u (which means that u will be printed).

This still does not mean that the primes (all of which will be printed) are printed in order. To get that we need to have

$$\models \forall u[(\text{at } \lambda_1 \wedge y_1 = u) \rightarrow \Box(\text{at } \lambda_0 \rightarrow y_1 > u)]$$

This simple example does, we hope, show that we can use a modal logic to state (and then to prove) properties of programs. In this simple case the only apparent benefit seems to be that precise statements can be made relatively easily readable. You should contrast this with trying to express the above purely in first-order logic, that is with the sequentiality explicitly mentioned. However, the real power of using this logic becomes apparent if we consider describing computations involving several programs which run simultaneously and interacting with one another. As an example, we might express what is known as the 'liveness' of a program, that is the property that the program never 'gets stuck' (perhaps waiting for a communicating program to give it some input) indefinitely as

$$\models \text{at } \lambda \rightarrow \Diamond \neg \text{ at } \lambda$$

that is if the program is at a particular location then there is an accessible state in which it has moved from that location. Many other such important properties can be so expressed.

8.9 Tableaux for intuitionistic logic

Computationally, the assumption that, for any proposition S, $S \vee \neg S$ is valid leads to problems. For example consider the program:

```
select
    S:print "hello"
   ¬S:print "hello"
endselect
```

This can be shown to be equal to the program

```
print "hello"
```

using the rule

```
if P = Q then
   select ... Ci: P ... Cj: Q ... endselect
= select ... Ci ∨ Cj:P endselect
```

as follows:

```
select
     S: print "hello"
   ¬S: print "hello"
endselect
= select S ∨ ¬S : print "hello" endselect
= select true : print "hello" endselect
= print "hello"
```

However, if S is a proposition whose truth-value is not known, for instance if S is 'there is an infinite number of twin primes', then the original program, when executed, gets no further than the test. Thus, the proof above must be wrong when compared with the execution.

Considerations like this may lead us to think about logics where the assumption that, for any $\mathscr{S}$, $\mathscr{S} \vee \neg \mathscr{S}$ is valid is not made.

We first looked at such a logic in Chapter 5 where we gave a natural deduction system for it. We shall now give a tableau system too, but first we recap a little.

Intuitionistic logic rejects this basic assumption of classical logic that every sentence is either true or false, regardless of whether or not we know which it is. Intuitionistic logic requires that if we say something is true (false) then we have shown that it is true (false) by, for instance, constructing a proof or disproof. So, given an arbitrary sentence, we cannot say it is true or false since in general we will have shown neither.

A slogan of intuitionistic logic is 'provability equals truth'; that is, speaking intuitionistically, if you say a sentence is true then you mean the sentence has been proved. Thus, to say a sentence is either true or false is to mean that it has either been proved or disproved, but this is clearly not true of every sentence. So, while classically $\mathscr{S} \vee \neg \mathscr{S}$ (where $\mathscr{S}$ is any sentence) is a tautology, intuitionistically it certainly is not. Thus, we reject the so-called 'law of the excluded middle'. We shall have more to say on all of this in the next chapter.

We next want to develop a tableau procedure for deciding the truth of a sentence under its intuitionistic meaning. It turns out that we can already do this by using the procedure above for S4. What we do is to model intuitionistic reasoning within a certain modal logic, and S4 turns out to be the one that works.

First of all, recall that an S4 logic is one in which the accessibility relation is reflexive and transitive. Now, if we consider **states of knowledge** to be our possible worlds then these clearly have a reflexive and transitive relation between them. That is, once in a given state of knowledge we assume that all the knowledge in that state is accessible to us (where by 'us' we mean 'ourselves as working mathematicians, scientists and so on'). Also, if state of knowledge A leads to state of knowledge B, and B leads to state of knowledge C then C is accessible from A too. This, we feel, is a reasonable model for the development of knowledge.

Now, the model becomes complete if we follow a certain translation. If $\mathscr{S}$ is an atomic sentence that is intuitionistically true, then it is classically provable. Furthermore, if it is provable in state of knowledge A then it is provable in all states of knowledge accessible from A (that is 'in the future' if we think of the accessibility as allowing access between a state and its future). That is, we assume that something once known is not 'not known'. If we formalize this we have that

A makes $\mathscr{S}$ intuitionistically true iff for any state of knowledge B accessible from A, B allows us to classically prove $\mathscr{S}$

Now, if you look back to the definition of $\Box$ you will see that the following is the case

$A \models_{S4} \Box\mathscr{S}$ iff for all possible worlds B such that B is accessible from A we have that $B \models_{S4} \mathscr{S}$.

Since 'allows us to prove classically' can be summed up as '$\models_{S4}$' and since our possible worlds are states of knowledge we have

A makes $\mathscr{S}$ intuitionistically true iff for all possible worlds B accessible from A we have $B \models_{S4} \mathscr{S}$

and further

A makes $\mathscr{S}$ intuitionistically true iff $A \models_{S4} \Box\mathscr{S}$

which we write, finally, as

$A \models_I \mathscr{S}$ iff $A \models_{S4} \Box\mathscr{S}$.

Therefore, we can test the intuitionistic validity of $\mathscr{S}$ by testing the S4-validity of $\Box\mathscr{S}$. In general we have to extend this to other than just the atomic sentences. We have the following **translation function** $\boldsymbol{T}$:

$T(\mathscr{S}) = \Box\mathscr{S}$, for $\mathscr{S}$ atomic

$$T(\mathscr{S} \wedge \mathscr{T}) = \Box(T(\mathscr{S}) \wedge T(\mathscr{T}))$$
$$T(\mathscr{S} \vee \mathscr{T}) = \Box(T(\mathscr{S}) \vee T(\mathscr{T}))$$
$$T(\mathscr{S} \rightarrow \mathscr{T}) = \Box(T(\mathscr{S}) \rightarrow T(\mathscr{T}))$$
$$T(\mathscr{S} \leftrightarrow \mathscr{T}) = \Box(T(\mathscr{S}) \leftrightarrow T(\mathscr{T}))$$
$$T(\neg \mathscr{S}) = \Box \neg T(\mathscr{S})$$

which, for example, formalizes the idea that $\mathscr{S} \wedge \mathscr{T}$ is intuitionistically true if it is provable that $\mathscr{S}$ is intuitionistically true and $\mathscr{T}$ is intuitionistically true. So, if we have $A \wedge B$, where A and B are atomic, then we have that $A \wedge B$ is intuitionistically true iff A is provable and B is provable, as we should have expected. Also we have that $\mathscr{S} \vee \mathscr{T}$ is intuitionistically true iff $\mathscr{S}$ is intuitionistically true or $\mathscr{T}$ is intuitionistically true (which makes it clear that we should not expect $\mathscr{S} \vee \neg \mathscr{S}$ in general). The last clause says that $\neg \mathscr{S}$ is true is read as $\Box \neg \Box \mathscr{S}$, that is (approximately) it is provable that $\mathscr{S}$ is not provable. The other translations can be read similarly.

We then have the property

$$\models_I \mathscr{S} \text{ iff } \models_{S4} T(\mathscr{S})$$

for $\mathscr{S}$ any sentence.

Thus, to test the validity of a sentence intuitionistically, we first translate it using T and then test the validity of the resulting sentence by using the procedure for S4.

For example, the sentence $A \vee \neg A$ gives

$$T(A \vee \neg A) =$$
$$\Box(T(A) \vee T(\neg A)) =$$
$$\Box(\Box A \vee \Box \neg T(A)) =$$
$$\Box(\Box A \vee \Box \neg \Box A)$$

Then, the S4 procedure will show that this sentence is not valid, as we did in Section 8.7, and hence $A \vee \neg A$ is not valid intuitionistically.

Exercise 8.5

(a) Decide on the classical and intuitionistic validity of each of the following sentences:

(i) $A \vee \neg A$

(ii) $\neg \neg A \rightarrow A$

(iii) $A \rightarrow \neg \neg A$

(b) We can extend the translation function T above to cope with the quantifiers as well. T is extended so that as well as the rules above we have:

$$T(\forall x \mathscr{S}) = \forall x \Box T(\mathscr{S})$$
$$T(\exists x \mathscr{S}) = \exists x \Box T(\mathscr{S})$$

using this, decide on the classical and intuitionistic validity of:

(i) $\forall x A(x) \rightarrow \exists x A(x)$

(ii) $\neg \forall x \neg A(x) \rightarrow \exists x A(x)$

SUMMARY

- Predicate calculus can only formalize types of reasoning in which sentences are true or false timelessly whereas in computer science, particularly, we often want to formalize systems in which propositions are true in some situations and false in others (such propositions are called *contingent*). To do this we introduced into the language two new *modalities*, $\Diamond \mathscr{S}$ which is read 'possibly $\mathscr{S}$' and whose intended interpretation is that $\Diamond \mathscr{S}$ is true if $\mathscr{S}$ is true in *some* possible situation, and $\Box \mathscr{S}$ which is read 'necessarily $\mathscr{S}$' and is true if $\mathscr{S}$ is true in *all* possible situations.
- These extensions to the language are given a semantics by introducing the notion of a *frame*. A frame consists of a set of *possible worlds* together with a relation of *accessibility* between worlds. Propositions may be true in some worlds and false in others. If $\mathscr{S}$ is a sentence then $\Box \mathscr{S}$ is true in a particular world if $\mathscr{S}$ is true in *all* worlds accessible from that world, and $\Diamond \mathscr{S}$ is true if $\mathscr{S}$ is true in *some* world accessible from it. Thus modal logic is not truth-functional. The truth-value of a formula such as $\Box \mathscr{S}$ or $\Diamond \mathscr{S}$ cannot be calculated from the truth-value of $\mathscr{S}$ at that world alone. The truth-value of the other logical connectives, however, depends only on their value at the world in which it is calculated, and so they are truth-functional and have the same meaning as in predicate calculus.
- A model in modal logic consists of a frame and an interpretation that specifies which atomic formulas are *forced* (true) in each possible world. A formula is satisfiable if there is an interpretation and a world in which it is forced. The validity of a formula can be disproved by displaying a counter model in the form of a *possible worlds diagram*, showing the frame as a directed graph together with the propositions that are forced at each world (node of the graph).
- Many systems of modal logic can be defined, depending on the properties of the accessibility relation. For example, the modal logic called S4 is the system in which the accessibility relation is reflexive and transitive. All such systems, however,

have the tautologies of propositional logic together with some basic equivalences such as $\Diamond\mathscr{S} \leftrightarrow \neg\Box\neg\mathscr{S}$. For different accessibility relations in the semantics there are corresponding proof systems. We have shown how a tableau system can be defined for S4.

- Modal logic can be used to give a semantics for intuitionistic logic, the logic that is obtained by dropping the Contradiction rule from the natural deduction system for predicate calculus. We saw in Chapter 5 that a simple extension of the truth-table idea to many linearly-ordered truth-values is not sufficient. However, a possible-worlds semantics can be given for intuitionistic logic. The appropriate accessibility relation is reflexive and transitive and we show how the tableau system for S4 can be adapted to yield a proof procedure. An atomic sentence $\mathscr{S}$ is intuitionistically valid iff $\Box\mathscr{S}$ is valid in S4.

Further Study

9.1 Introduction

This chapter is intended as an appetizer for further study. Having worked through the book the reader will, we hope, be in a position to go on to more advanced work: to look at logics with more sophisticated bases; to look further, and more sceptically, at the philosophical basis for much that we have presented here; to see how logics have been developed for talking about, and making progress towards solutions for, many problems in computer science.

We shall, in what follows, do no more than briefly mention some problems and further work based on them, and direct the reader to some reasonably accessible texts that will allow further study. By no means do we claim to present a complete bibliography, merely one that will allow the reader to move on to the next stage in each area.

9.2 Connection method

9.2.1 Basic ideas

The connection method was introduced by Bibel (1987) as an efficient method for deciding validity in predicate logic. It turns out that it has very close connections with the tableau method that we looked at in Chapter 4, and can be seen, especially in its later developments, as a very efficient way of implementing tableau systems. Indeed, this has been demonstrated best by Wallen (1987) not only for classical logics but non-standard ones too.

We shall give here just the first few steps towards the full connection method. It is rather complicated to describe as the logical and algorithmic aspects that give it its efficiency tend to be rather intertwined.

Consider the usual first example of a predicate logic argument

A1: $\forall x(\text{Man}(x) \rightarrow \text{Mortal}(x))$
A2: Man(Socrates)
TH: Mortal(Socrates)

As we know, we can express this argument as

$$(\forall x(\text{Man}(x) \rightarrow \text{Mortal}(x)) \wedge \text{Man}(\text{Socrates})) \rightarrow \text{Mortal}(\text{Socrates}) \quad \textbf{(9.1)}$$

If we then replace $\rightarrow$ by its expression in terms of $\vee$ and $\neg$ we get

$$\begin{aligned} &\neg[\forall x(\text{Man}(x) \rightarrow \text{Mortal}(x)) \wedge \text{Man}(\text{Socrates})] \vee \text{Mortal}(\text{Socrates}) \\ \leftrightarrow &\neg[\forall x(\neg\text{Man}(x) \vee \text{Mortal}(x)) \wedge \text{Man}(\text{Socrates})] \vee \text{Mortal}(\text{Socrates}) \\ \leftrightarrow &\exists x(\text{Man}(x) \wedge \neg\text{Mortal}(x)) \vee \neg\text{Man}(\text{Socrates}) \vee \text{Mortal}(\text{Socrates}) \end{aligned} \quad \textbf{(9.2)}$$

The next step is to write sentence 9.2 in the form of a **matrix**. This is a two-dimensional representation of any sentence that is in the form of a disjunction of conjuncts, the **disjunctive normal form**, such as sentence 9.2. So the matrix here is

$$\begin{bmatrix} \text{Man}(x) & & \\ & \neg\text{Man}(\text{Socrates}) & \text{Mortal}(\text{Socrates}) \\ \neg\text{Mortal}(x) & & \end{bmatrix}$$

and by convention variables, such as x here, are existentially quantified. The above is said to be a matrix consisting of a set of **clauses**, which are in disjunction, and each clause consists of a set of **literals**, which are in conjunction. If we cross the matrix from left to right, visiting one literal from each clause, we form a **path**.

A pair of literals with the same predicate symbol, where one of the literals is negated, is called a **connection**. A set of connections is called **spanning** if each path through a matrix contains a connection from the set.

In the matrix above there are two paths

(p1) $\{\text{Man}(x), \neg\text{Man}(\text{Socrates}), \text{Mortal}(\text{Socrates})\}$
(p2) $\{\neg\text{Mortal}(x), \neg\text{Man}(\text{Socrates}), \text{Mortal}(\text{Socrates})\}$

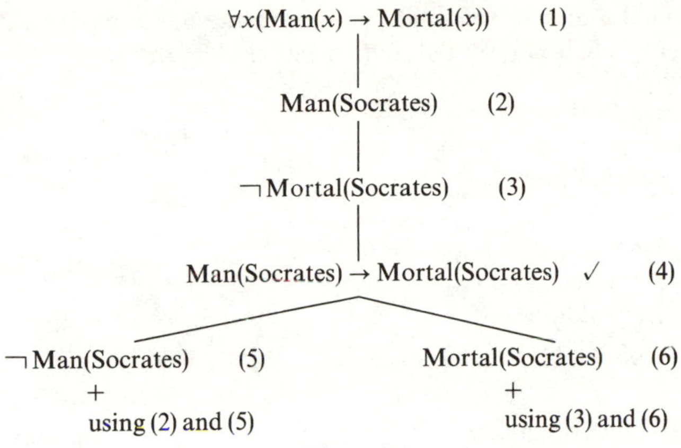

Figure 9.1.

and two connections

(c1) $\{\text{Man}(x), \neg\text{Man}(\text{Socrates})\}$

(c2) $\{\neg\text{Mortal}(x), \text{Mortal}(\text{Socrates})\}$

Since each of p1 and p2 contains one of c1 or c2, that is c1 $\subset$ p1 and c2 $\subset$ p2, then by definition {c1, c2} is a spanning set of connections for the matrix.

It can be shown that a sentence is valid iff each element of a spanning set of connections for the matrix form of S can be made into a complementary pair of literals, that is the literals have the same predicate letter, one negated, and the same arguments.

In the case above c1 and c2 can each be made complementary by the substitution of Socrates for x. Therefore, we conclude that the original sentence 9.1 is valid.

It is illuminating to compare this with the tableau proof (Figure 9.1):

Note that we have made the same substitution and, essentially, the same connections, shown here by contradictions. We can think of this set of contradictions as spanning since they close every path of the tableau. Note, further, that by listing the atomic sentences of all paths on the tableau we regain all of the paths through the matrix.

9.2.2 Generating spanning sets

Consider the matrix shown below:

$$\begin{array}{ccccc} 0 & 1 & 2 & 3 & 4 \end{array}$$
$$\left[\begin{array}{ccccc} L & \neg L & \neg M & K & L \\ & & & & \\ \neg K & M & & & M \end{array}\right]\begin{array}{c} 0 \\ \\ 1 \end{array}$$

We have added a grid to this matrix to make the naming of (positions of) literals easier in what follows. This matrix has eight paths:

{00, 01, 02, 03, 04}	p1
{00, 01, 02, 03, 14}	p2
{00, 11, 02, 03, 04}	p3
{00, 11, 02, 03, 14}	p4
{10, 01, 02, 03, 04}	p5
{10, 01, 02, 03, 14}	p6
{10, 11, 02, 03, 04}	p7
{10, 11, 02, 03, 14}	p8

The task is to find a spanning set of connections, that is a connection for each path.

We choose a clause, say 0, and a literal within it, say 00. We then look for a complementary literal in some other clause, say 1, so we find the literal 01. After making this connection we can form three disjoint sets of paths:

(a) those containing the connection {00, 01} – these need no more work

(b) those still to be worked on which contain 00

(c) those still to be worked on which do not contain 00

so here we have

(a) {p1, p2}

(b) {p3, p4}

(c) {p5, p6, p7, p8}

Next we put set (c) aside, perhaps on a stack in an implementation of this procedure, and look at set (b). With (b) we now repeat the procedure above. We choose a clause, say 1, and a literal not yet in a connection (here there is only one, namely 11). Then, we search the other clauses to form a connection with 11. Say we find 02. Then, we partition the set (b), as before, into three disjoint subsets:

(ba) {p3, p4}

(bb) { }

(bc) { }

Clearly, this subproblem is solved since there are no clauses in the 'still to be connected' sets. So, we go back to our stack of sets containing paths yet to have connections found for them and work on them. Here, we work on set (c). This gives (ca) {p5, p6, p7, p8}

(cb) { }

(cc) { }

by choosing the connection {10, 03}. Now, since there are no more problems stacked the whole problem is solved and we have our spanning set. Hence, the initial matrix represents a valid sentence.

There is, of course, much more to say about the connection method. In particular we have not mentioned here how we deal with first-order sentences. Having generated spanning sets using the above mechanism we would need to find substitutions that make the literals in the sets complementary. This involves the use of unification and also some mechanism for respecting the restrictions that govern universal and existential quantifiers. We have also not said anything about the algorithms developed by Bibel for efficiently implementing the method. However, all these questions really take us too far into the research area of automated reasoning and so too far outside the scope of the current text. To treat the method properly would take a whole textbook, as Bibel indeed does (Bibel, 1987), so we prefer to leave the topic here and let readers follow up if they wish.

9.3 LCF

9.3.1 Introduction

The name 'LCF' (logic for computable functions) nowadays covers more than one system. The one that we shall look at here, which is the original one, is more properly called 'Edinburgh LCF'. It grew out of work that was started at Stanford University in California, USA. It is essentially a computer program that acts as a proof checker for a logic of computable functions.

The interactive nature of this system means that instead of a problem being stated and then given to the program to work on until it gets an answer (or, more usually, until it runs out of storage or the user runs out of patience), the idea of a proof checker (as opposed to a theorem prover) is that the user issues commands to build proofs step by step in an interactive manner. So, a step might be, 'apply modus ponens to theorems *A* and *B*'. The program will check that the premises of *A* and *B* are of the right form for this step to be correct and then go on to produce the conclusion. (Our SML program for proof checking does just this, though not interactively.)

In the case of Edinburgh LCF the way of issuing commands has been embodied in a language that has an expressive functional subset. This language is ML, for meta-language, and is the precursor to the language SML that we have been using in this book.

Using a language with the expressiveness of ML means that not only can proof rules be encoded but also rules for composing proof rules together. This feature is a very important one.

The form of proof used in LCF is that of natural deduction. Each rule of inference in the logic is expressed as an ML function which has as result a value of the ML abstract type thm, that is a theorem. There are also functions which work on the syntax of the language to break up sentences and build new ones, rather in the manner of our functions like free_for, free_in or instance_of.

9.3.2 An example proof

As an example of how the system is used we shall re-express a proof given above in Chapter 5. This theorem to be proved is

$$\vdash p \rightarrow (q \rightarrow (p \wedge q))$$

and, adapting the proof from Chapter 5, we have the following proof:

$$\dfrac{\dfrac{\dfrac{\dfrac{}{\{p\} \vdash p}\text{ ASSUME} \qquad \dfrac{}{\{q\} \vdash q}\text{ ASSUME}}{\{p, q\} \vdash p \wedge q}\text{ CONJ}}{\{p\} \vdash q \rightarrow (p \wedge q)}\text{ DISCH}}{\vdash p \rightarrow (q \rightarrow (p \wedge q))}\text{ DISCH}$$

Having discovered a proof like the above, we now need to use the LCF system to check that it is in fact a correct proof. As we said before, the rules of the logic are implemented as ML functions. These functions check that the premises given to each application of a rule really do meet the requirements given in the rule definition. Further, the only way of getting values of type thm, that is theorems, in the implementation is as the result of the functions which implement the rules. Hence, if something is of type thm then it really is a theorem. So, if we can show that at each stage of the proof above the statements are in thm then we know that the proof is correct, that is we will successfully have checked it.

The first rule we need to use is

```
ASSUME: form -> thm
```

where the formula w is mapped to $\{w\} \vdash w$. So, if we are sitting with the system waiting in readiness for us to interact with it, we type

```
let a1 = ASSUME "p";;
let a2 = ASSUME "q";;
```

and get the response

```
val a1 = ["p"]]-  "p" : thm
val a2 = ["q"]]-  "q" : thm
```

(where]– is the turnstile). Then we use the function

```
CONJ: (thm × thm) -> thm
```

which takes a theorem of the form $A_1 \vdash w_1$ and one of the form $A_2 \vdash w_2$ and returns $A_1 \cup A_2 \vdash w_1 \wedge w_2$ so here we type

```
let a3 = CONJ(a1,a2);;
```

which gives

```
val a3 = ["p","q"]]- "p & q" : thm
```

Then we use

```
DISCH: form -> thm -> thm
```

where w and $A \vdash w'$ are mapped to $A' \vdash w \rightarrow w'$ where A' is A with all occurrences of w deleted. So here we type

```
let a4 = DISCH "q" a3;;
```

to get

```
val a4 = ["p"]]- "q imp (p & q)" : thm
```

and finally

```
let a5 = DISCH "p" a4;;
```

to get

```
val a5 = ]- "p imp (q imp (p & q))" : thm
```

as required. So, our proof has been checked and found correct. If we had actually made a mistake in any of the above steps then the system would have informed us of this and disallowed the step. That is, a step is only allowed if the rule is being used correctly. Any misuses are reported and the user has to adapt the alleged proof so that all its steps are correct uses of the rules.

9.3.3 Tactics

Having seen how the system might be used on a very simple example where the user decides at each step which rule to use, we now go on to look at a way in which we can begin to make the system do more work for us. This will depend on using the expressiveness of ML to program some more or less general-purpose problem-solving strategies.

One common way of constructing proofs in a natural deduction style (though it is not of course restricted to this style) is as described for natural deduction in Chapter 5, that is we use the rules of inference in a backward sense. That is, given the correct goal, we find a rule whose conclusion can be made to match the goal and hence, by reading off the associated premises, find some subgoals which are simpler to prove than the original goal. We keep on using this basic strategy until all the subgoals reduce to axioms or to sentences that we have proved previously.

For example, we can try to prove

$$\vdash (p \wedge q) \rightarrow (q \wedge p) \tag{9.3}$$

First we look for a rule whose conclusion matches the sentence. In general there will be many matches, so a searching mechanism will be required to make sure that we try all possibilities. In this case it turns out that two rules are applicable; the rule ASSUME or the rule DISCH as given above.

Clearly ASSUME gets us nowhere with our current goal so we use DISCH. We match with sentence 9.3 so that

$(p \wedge q)$ is w
$(q \wedge p)$ is w'
$\{\ \}$ is A
$\{\ \}$ is A'

and end up with the subgoal, that is the instantiated premise of DISCH, being

$$(p \wedge q) \vdash (q \wedge p) \tag{9.4}$$

We also need to record, for use later, the reason why this is a legitimate step. In this case it is because of the rule DISCH. Now, we do the same with subgoal 9.4. The only rule that matches is CONJ which means we now have the subgoals

$$(p \wedge q) \vdash q \tag{9.5a}$$
$$(p \wedge q) \vdash p \tag{9.5b}$$

and we record CONJ as the reason for these subgoals. Tackling subgoal 9.5a first, the only rule which matches it is

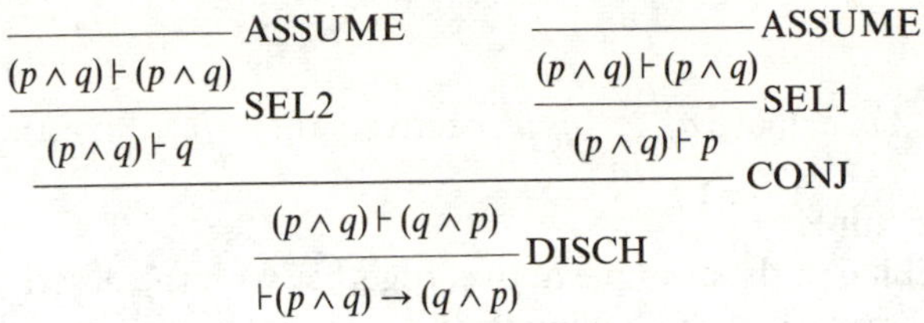

$$\dfrac{\dfrac{\overline{(p \wedge q) \vdash (p \wedge q)}\ \text{ASSUME}}{(p \wedge q) \vdash q}\ \text{SEL2} \qquad \dfrac{\overline{(p \wedge q) \vdash (p \wedge q)}\ \text{ASSUME}}{(p \wedge q) \vdash p}\ \text{SEL1}}{\dfrac{(p \wedge q) \vdash (q \wedge p)}{\vdash (p \wedge q) \rightarrow (q \wedge p)}\ \text{DISCH}}\ \text{CONJ}$$

Figure 9.2.

SEL2: thm –> thm — takes a theorem of the form $A \vdash w_1 \wedge w_2$ and returns $A \vdash w_2$

to give the goal

$$(p \wedge q) \vdash (p \wedge q) \tag{9.6a}$$

with the reason SEL2. Similarly, subgoal 9.5b needs SEL1

SEL1: thm –> thm — takes a theorem of the form $A \vdash w_1 \wedge w_2$ and returns $A \vdash w_1$

to give

$$(p \wedge q) \vdash (p \wedge q) \tag{9.6b}$$

with reason SEL1. Then goals 9.6a and 9.6b can be satisfied by ASSUME applied to $(p \wedge q)$, and leading to no further subgoals.

Now, we can put all the validations together to get

```
DISCH(CONJ(SEL2(ASSUME(p&q)),SEL1(ASSUME(p&q))))
```
(9.7)

which can be thought of a linear way of describing what, in Chapter 5, would have been more conventionally written as shown in Figure 9.2.

The task of mapping a goal to a list of subgoals and a validation is carried out by what, in LCF parlance, are called **tactics**. (Here we are going to present a simplified version of the full LCF idea. It will, however, convey the main points about use of tactics.) A **validation** is some action which makes a tactic T **valid**. This means that

if g and $g_1, \ldots, g_n$ are goals and v is a validation and $T(g) = ([g_1, \ldots, g_n], v)$

then

if $v_1, \ldots, v_n$ achieve $g_1, \ldots, g_n$ respectively then $v(v_1, \ldots, v_n)$ achieves g

which rule we call V.

By looking at the example above, together with the descriptions of the rules, we can assemble the following tactics, subgoal lists and validations:

DISCHtac	where $(A', w \rightarrow w')$ is mapped to $([(A \cup \{w\}, w')], \lambda x . (\text{DISCH}\ w\ x))$
CONJtac	where $(A_1 \cup A_2, w_1 \wedge w_2)$ is mapped to $([(A_1, w_1), (A_2, w_2)], \lambda(x, y) . \text{CONJ}(x, y))$
SEL1tac	where (A, w_1) and w_2 are mapped to $([(A, w_1 \wedge w_2)], \lambda x . (\text{SEL1}\ x))$
SEL2tac	where (A, w_2) and w_1 are mapped to $([(A, w_1 \wedge w_2)], \lambda x . (\text{SEL2}\ x))$
ASSUMEtac	where $(\{w\}, w)$ is mapped to $([\ ], \text{ASSUME}\ w)$

and note that this last tactic is a special case since it produces no subgoals.

Now we can re-trace our steps in the construction of the proof above. First, at step 1, we apply DISCHtac to '$(p \& q)$ imp $(q \& p)$' which gives us

$$([([(p \& q)], (q \& p))], \lambda x . (\text{DISCH}(p \& q)\ x))$$

(and here we have left out some of the quotes and parentheses to make the text a little easier to read). Then we begin on the subgoal list, which in this case means dealing with just one subgoal. We use CONJtac to get, at step 2

$$([([(p \& q)], q), ([(p \& q)], p)], \lambda(x, y) . \text{CONJ}(x, y))$$

and then SEL2tac on the first subgoal and SEL1tac on the second to get, at steps 3 and 4

$$([([(p \& q)], (p \& q))], \lambda x . (\text{SEL2}\ x))$$
$$([([(p \& q)], (p \& q))], \lambda x . (\text{SEL1}\ x))$$

and each of these is solved with ASSUMEtac, at steps 5 and 6, to give, for both cases

$$([\ ], \text{ASSUME}\,(p \& q))$$

which has no subgoals, so we are finished. Now we just need to collect up the validations in each case and use the rule V for achieving goals. So, to achieve the two subgoals resulting from steps 3 and 4 we apply ASSUME $(p \& q)$ to what achieves the subgoals at steps 5 and 6, but since the list is empty in each case, that is there are no subgoals, we apply it to nothing. Then, to achieve the first subgoal at step 2 we apply $\lambda x.(\mathsf{SEL2}\ x)$ to $\mathsf{ASSUME}(p \,\&\, q)$ (according to rule V) and to achieve the second subgoal at step 2 we apply $\lambda x.(\mathsf{SEL1}\ x)$ to $\mathsf{ASSUME}(p \,\&\, q)$. To achieve the subgoal at step 1 we apply $\lambda(x, y).\mathsf{CONJ}(x, y)$ to $(\lambda x.(\mathsf{SEL2}\ x)\ \mathsf{ASSUME}(p \,\&\, q),\ \lambda x.(\mathsf{SEL1}\ x)\ \mathsf{ASSUME}(p \,\&\, q))$. Finally, to get the original goal we apply $\lambda x.(\mathsf{DISCH}(p \,\&\, q)x)$ to $(\lambda(x, y).\mathsf{CONJ}(x, y)\ (\lambda x.(\mathsf{SEL2}\ x)\ \mathsf{ASSUME}(p \,\&\, q), \lambda x.(\mathsf{SEL1}\ x)\ \mathsf{ASSUME}(p \,\&\, q)))$ to get

$$\lambda x.(\mathsf{DISCH}\,(p \,\&\, q)x)\,(\lambda(x, y).\mathsf{CONJ}(x, y)(\lambda x.(\mathsf{SEL2}\ x)\,\mathsf{ASSUME}\,(p \,\&\, q),\\ \lambda x.(\mathsf{SEL1}\ x)\,\mathsf{ASSUME}\,(p \,\&\, q)))$$

which simplifies, by the rules of the λ-calculus, to the expression 9.7.

So, we can see that by having tactics, which are essentially the rules of inference backwards with extra mechanism for book-keeping purposes, we can formalize the idea of working backwards from goal to simpler and simpler subgoals. The next stage, which we will not go into, allows these tactics to be combined together, by functions called **tacticals**, by operators to form more and more complex tactics to get nearer and nearer to the aim of making more and more of the proof construction process automatic, at least in fairly uniform and simple cases.

We will leave our presentation of LCF here, but fuller details can be found in Gordon *et al.* (1979).

9.4 Temporal and dynamic logics

9.4.1 Temporal logic

One particularly important use for the ideas of modal logics are logics where the modalities are related to the time at which or for which statements are true or false. We looked briefly at an approximation to this idea when we showed an example of using modal logics to describe computations in Chapter 8. There the possible worlds were states in a computation and the accessibility relation related worlds which were linked by the execution of a program statement. This can be taken further and instead of states in the computation being linked by program statements we can think of moments in time linked by the passage of time. Under this more general presentation the accessibility relation relates states that are temporally earlier with states that are temporally later. Finally, this idea can then be used to model any system where there is a temporal relation between states.

Let us set up a simple temporal logic, as we call logics which reason about time (and they are also called 'tense logics' by other authors). We follow the pattern set when we talked about modal logics, so first we need a frame.

We let P be a set of possible worlds and R be an accessibility relation, but this time, to remind ourselves that we are dealing with an 'earlier–later than' relation we shall write R as $<$. As suggested by the use of this symbol, we shall make $<$ transitive. We shall also make $<$ **connected**. This means that no time that we can get to is isolated from past or future times. These two conditions mean that the sort of time we shall look at is the usual linear one that is used in everyday reasoning (though it does break down once we start to think of relativistic effects). The conditions can be formalized as

$<$ is connected: $\forall t \forall t'(t < t' \vee t = t' \vee t' < t)$
$<$ is transitive: $\forall t \forall t'((t < t' \wedge t' < t'') \rightarrow t < t'')$

Now, following our pattern for modal logics we can introduce the counterpart to $\Box$ and $\Diamond$. Following our previous use of the notation we will write

$t \Vdash \mathcal{S}$

to mean that $\mathcal{S}$ is true at time t.

There are usually taken to be four **temporal modalities** and these are as follows, where $\mathcal{S}$ is any statement:

$F\mathcal{S}$	there is a future time when $\mathcal{S}$ will be true
$G\mathcal{S}$	at all future times $\mathcal{S}$ will be true
$P\mathcal{S}$	there is a past time when $\mathcal{S}$ was true
$H\mathcal{S}$	at all past times $\mathcal{S}$ was true

These can be formalized as:

$t \Vdash_v F\mathcal{S}$	iff	there is a t' such that $t < t'$ and $t' \Vdash_v \mathcal{S}$
$t \Vdash_v G\mathcal{S}$	iff	for all t', if $t < t'$ then $t' \Vdash_v \mathcal{S}$
$t \Vdash_v P\mathcal{S}$	iff	there is a t' such that $t' < t$ and $t' \Vdash_v \mathcal{S}$
$t \Vdash_v H\mathcal{S}$	iff	for all t', if $t' < t$ then $t' \Vdash_v \mathcal{S}$

with the definitions for the non-temporal operators as before. These definitions give, analogous to the case with $\Box$ and $\Diamond$,

$\neg F\mathcal{S} \leftrightarrow G\neg\mathcal{S}$
$\neg P\mathcal{S} \leftrightarrow H\neg\mathcal{S}$

It turns out that for this system there is also a tableau method which can be used to decide validity, just as in the modal and classical cases.

This, of course, is just the beginning of the story. We can go on to ask whether time is discrete, that is broken into indivisible moments, or continuous, in which case, like the real line, there is always a moment of time

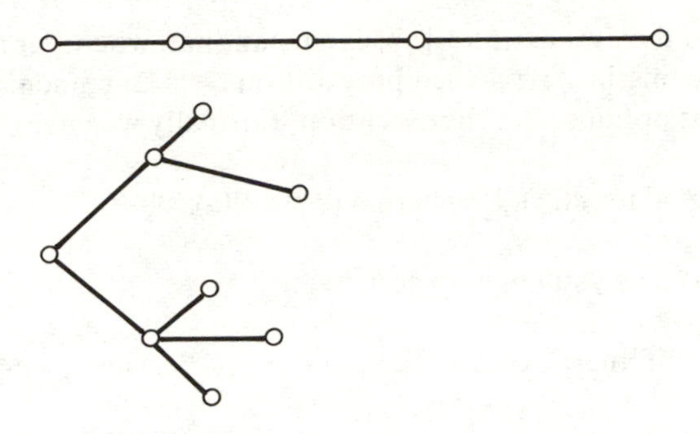

Figure 9.3.

between any two moments. Also, we can model the idea that the future is not linear, as we have it here, but branching, in the sense that, even though a given observer passes through a sequence of moments, the future might be a branching structure of possibilities through which an individual traces a single, linear path. The difference can be suggested by comparing the diagrams in Figure 9.3, for linear and branching time respectively.

We have here only looked at a tiny fragment of the huge and varied philosophical and mathematical literature on temporal logics. For further study we refer the reader to Rescher and Urquhart (1971). For work on temporal logics directly related to programming Manna and Pnueli (1981) is a good starting point.

9.4.2 Dynamic logic

This is another example of an extension of the basic ideas of modal logic. If we take P, the set of possible worlds, to be possible states of a computation then the accessibility relation R will relate a state s to a state t. However, we want the fact that states are related, because execution of a program causes transition from one to the next, to be reflected in the accessibility relation. The **dynamic logic** way of doing this is to index the relation with the program. So, if s and t are states of a computation and we get from s to t by executing the program α then we should have that $(s,t) \in R(\alpha)$. That is, the relation R changes depending on what program we are currently dealing with. This is clearly an extension of the idea of accessibility relation that we saw previously.

Now we want to see how this change is reflected at the level of propositions. Recall the definition of $\Box$:

$$s \Vdash \Box \mathscr{S} \text{ iff for all } t \in P \text{ such that } (s,t) \in R, t \Vdash \mathscr{S}$$

Since the relation R will now depend upon the program α we make the necessity operator depend upon it too. If the frame and valuation (as before) are understood then we write $s \Vdash [\alpha]\mathscr{S}$ to mean that, whenever the program α is executed starting in state s, then proposition $\mathscr{S}$ is true under the valuation in the state that obtains after the execution. Formally we have

$$s \Vdash [\alpha]\mathscr{S} \text{ iff for all } t \in P \text{ such that } (s, t) \in R(\alpha),\ t \Vdash \mathscr{S} \qquad \textbf{(9.8)}$$

and we also have, as you might expect,

$$s \Vdash \langle\alpha\rangle\mathscr{S} \text{ iff there exists a } t \text{ such that } (s, t) \in R(\alpha) \text{ and } t \Vdash \mathscr{S} \qquad \textbf{(9.9)}$$

that is $s \Vdash \langle\alpha\rangle\mathscr{S}$ means that if α is executed in state $\mathscr{S}$ then there is a state that the computation will be in where $\mathscr{S}$ obtains. As before, a sentence is valid in a frame iff it is true for all valuations in the frame and a sentence is valid iff it is valid in all frames.

The language of propositions, with the extension of indexing the modalities, is just as usual. The language of programs is usually based, using the initial work in Pratt (1976), on the following primitives:

do α then do β	$\alpha; \beta$
do either α or β	$\alpha \cup \beta$
do α a finite number of times	α^*
evaluate the proposition $\mathscr{S}$, if it is true then carry on with the evaluation, otherwise fail.	$\mathscr{S}?$

The meaning of these programs is given in an obvious way. So, for instance, we have

$$\text{if } (s, t) \in R(\alpha) \text{ and } (t, r) \in R(\beta) \text{ then } (s, r) \in R(\alpha; \beta), \text{ for any } s, r, t \in P$$

Using these program forms we can introduce more expressive statements by definition. For example we have

while $\mathscr{S}$ do α	is defined to be	$(\mathscr{S}?; \alpha)^*; \neg\mathscr{S}?$
if $\mathscr{S}$ then α else β	is defined to be	$(\mathscr{S}?; \alpha) \cup (\neg\mathscr{S}?; \beta)$

Using this simple propositional language we can make statements about the correctness of algorithms. For instance, the statement

$$(n \geqslant 0) \rightarrow [x, y := n, 1; \text{while } x > 0 \text{ do } (y := y * x; x := x - 1)]\,(y = n!)$$

is true if, whenever we start in a state where $n \geqslant 0$ (where a state is a tuple of variable values) and then execute the program, the final state will be one where $y = n!$, that is the statement says that the program correctly calculates the factorial function of n.

We can, more abstractly, also make statements about relations between programs. So, for instance, if $\mathscr{S}$ is any statement then if we can prove

$$\langle\alpha\rangle\mathscr{S} \leftrightarrow \langle\beta\rangle\mathscr{S}$$

then we can conclude that the programs α and β are equivalent, that is they give the same result when started in the same state.

We can also prove derived properties of programs that can serve to make proofs easier. For instance, under the intended interpretation, it is clear that we would expect that if

$$\Vdash \mathscr{S} \rightarrow [\alpha]\mathscr{T} \text{ and } \Vdash \mathscr{T} \rightarrow [\beta]\mathscr{R}$$

then

$$\Vdash \mathscr{S} \rightarrow [\alpha;\beta]\mathscr{R}$$

and we can show this as follows. By assumption, for any frame $\langle P, R\rangle$ and any valuation in that frame we have, for any $s \in P$, $s \Vdash \mathscr{S} \rightarrow [\alpha]\mathscr{T}$, that is if $s \Vdash \mathscr{S}$ then $s \Vdash [\alpha]\mathscr{T}$. So, for all $t \in P$ such that $(s, t) \in R(\alpha)$, $t \Vdash \mathscr{T}$. Similarly, again by assumption, we have that if, for any $t \in P$, $t \Vdash \mathscr{T}$, then for all $r \in P$ such that $(t, r) \in R(\beta)$, $r \Vdash \mathscr{R}$. So, putting this together, we have that if $s \Vdash \mathscr{S}$ then for all $t \in P$ such that $(s, t) \in R(\alpha)$ and for all $r \in P$ such that $(t, r) \in R(\beta)$, $r \Vdash \mathscr{R}$. By definition, this means that if $s \Vdash \mathscr{S}$ then, for all $s, r \in P$ such that $(s, r) \in R(\alpha;\beta)$, $r \Vdash \mathscr{R}$. And this means that if $s \Vdash \mathscr{S}$ then $s \Vdash [\alpha;\beta]\mathscr{R}$. This, finally, gives us the desired conclusion.

Here we have only given an idea of the expressiveness of dynamic logic and have given very little in the way of formal semantics. Its expressiveness can be extended by allowing the statements to be first order and then as well as quantifying over, say, numbers we can quantify over programs themselves. The formal semantics that we gave for modal logics can be extended to include a formal semantics for programs too. This in turn leads to an extension of the tableau system to provide algorithmic means for decision procedures and semidecision procedures as before. However, we shall leave this huge subject here since, again, we would need to write a whole book to do it justice.

9.5 Intuitionistic logic

9.5.1 Introduction

Intuitionistic logic was developed because mathematicians working at the turn of the century thought that the foundations of their subject were not secure. The basis for these thoughts came with problems in set theory like the Cantor paradox. This can be briefly described as follows. If A is a set then the size of the set is given by its **cardinal number** $\bar{\bar{A}}$, which is the set of all sets which have the same number of elements as A (or more precisely whose elements can be put into a one–one correspondence with those of A). Let $\bar{\bar{A}} \leqslant \bar{\bar{B}}$, for two sets A and B, mean that A has the same cardinality as a subset of B. Then, by a result of Cantor's, for any set A and its power set $P(A)$ we have $\bar{\bar{A}} < \overline{\overline{P(A)}}$. Now, let U be the set of all sets, that is the universal set. Then, by definition $P(U) \subseteq U$ so $\overline{\overline{P(U)}} \leqslant \bar{\bar{U}}$. But, by Cantor's result we have $\bar{\bar{U}} < \overline{\overline{P(U)}}$, which gives a contradiction. This is Cantor's paradox.

Frege and Hilbert, who thought that mathematics lacked proper foundations, set about trying to provide them.

Frege thought that the foundations should be built by basing everything on logic, as he considered this to be a more fundamental subject. Hilbert thought that mathematics should be built by setting up, say, arithmetic by using completely agreed upon and clear axioms and then using only completely clear 'meta-mathematical' arguments to construct all of the rest. Both these projects failed. Frege's ideas, as we said before, failed to form the foundations based on purely logical grounds because of the contradictions found by Russell. Hilbert's program, as it was called, failed because of the incompleteness theorems of Gödel. Brouwer started a third movement that tried to provide a proper foundation, and this movement is usually called **intuitionism**.

If you are working as an intuitionist then you want all of your mathematics to be **constructive**. Put simply, this means that you do your mathematics without use of the so-called law (or principle) of the excluded middle and without the use of non-constructive existence proofs.

The law of the excluded middle states that, for any statement $\mathcal{S}$, either $\mathcal{S}$ is true or $\neg\mathcal{S}$ is true. This is the fundamental assumption that we introduced when starting with classical logic. The general form of a non-constructive existence proof is one where we seek to prove $\exists x \mathcal{S}(x)$ by proving $\neg\forall x \neg\mathcal{S}(x)$.

As an example of the sort of proof that mathematicians do not want to give up, because of its elegance and simplicity and which uses the law of the excluded middle, we have the following time-honoured first example:

Theorem

There exist solutions of $x^y = z$ with x and y irrational and z rational.

Proof $\sqrt{2}$ is irrational. Also, $\sqrt{2}^{\sqrt{2}}$ is either rational or irrational. If $\sqrt{2}^{\sqrt{2}}$ is rational then put $x = \sqrt{2}$ and $y = \sqrt{2}$, then z is rational. If $\sqrt{2}^{\sqrt{2}}$ is irrational then put $x = \sqrt{2}^{\sqrt{2}}$ and $y = \sqrt{2}$, then $z = (\sqrt{2}^{\sqrt{2}})^{\sqrt{2}} = \sqrt{2}^2 = 2$ is rational. ■

Constructively the problem with this is that, having 'proved' the theorem is true, we still do not actually know how to give examples of the numbers x, y and z. This is rather unsatisfactory and certainly not constructive.

9.5.2 Brouwer's criticism

Brouwer said that mathematics was essentially something that people carried out in their minds. He saw mathematics as the activity of building mental models and finding relationships between them. Formalism, under his reading, is not a part of mathematics itself; however, since we like to communicate our mathematics to other people we need to express it in some precise language, hence the formalisms that have been developed.

Brouwer also saw as basic the idea of a unit and the idea of putting units together, from which we get the basic notion of counting and so the natural numbers. This analysis also immediately yields the idea of mathematical induction. This is in contrast to Frege's development of the natural numbers from purely logical principles.

However, Brouwer rejected the idea of the law of the excluded middle and the validity of non-constructive existence proofs. He believed that the problem that was at the root of the set theoretic paradoxes was that classical logic was generalized from the logic that is behind finite sets. This is what mathematics essentially dealt with until the work of Cantor, Dedekind and others, who introduced the idea of infinite sets as completed objects. It then seems as though people eventually forgot or ignored this development from the finite and used this actually rather restricted logic on sets in general. This misuse was what gave rise to the paradoxes.

9.5.3 Formalization of intuitionistic mathematics

The first formal system for the logical part of intuitionistic mathematics was presented in 1930 by Heyting. (A more readable, to a non-German reader, presentation is in Heyting (1956).) He gave an interpretation of the connectives which forms the basis of the system that we shall talk about below. The interpretation is as follows:

- $\mathscr{S} \wedge \mathscr{T}$ can be asserted iff both $\mathscr{S}$ and $\mathscr{T}$ can be asserted.
- $\mathscr{S} \vee \mathscr{T}$ can be asserted iff at least one of $\mathscr{S}$ and $\mathscr{T}$ can be asserted.
- $\neg\mathscr{S}$ can be asserted iff we have a method which, supposing we have constructed a proof of $\mathscr{S}$, would when applied to the proof for $\mathscr{S}$ lead to a contradiction.
- $\mathscr{S} \rightarrow \mathscr{T}$ can be asserted iff we have a method which when applied to a proof for $\mathscr{S}$ will give a proof for $\mathscr{T}$.
- $\forall x\mathscr{S}(x)$, with x ranging over $\mathscr{T}$, can be asserted iff we have a method

which, for any element a from $\mathcal{T}$, gives a method for constructing a proof of $\mathcal{S}(a)$.

- $\exists x.\mathcal{S}(x)$ can be asserted iff we can construct an element a and a proof of $\mathcal{S}(a)$.

Logical schemas, that is formulas with sentence variables, can be asserted iff we have a method which by specialization gives a method for constructing a proof for the formula with any propositions substituted for the variables. For example, we have $\mathcal{S} \to \mathcal{S}$ since the 'identity method' trivially can be specialized to provide a proof of this formula with any proposition in place of $\mathcal{S}$.

We can see how this explanation shows that the excluded middle, $\mathcal{S} \vee \neg\mathcal{S}$, is not valid. If it were valid then it would give a general method to solve any problem. That is, for any proposition $\mathcal{S}$ the general method would specialize to give either a proof of $\mathcal{S}$ or a proof of $\neg\mathcal{S}$. Since such a general method does not exist, the law is not valid.

Heyting actually gave his logic as a formal system. It consisted of ten axioms together with the usual rules of deduction, that is modus ponens and instantiation.

9.5.4 The link between proofs and programs

The connection between intuitionistic mathematics and programming comes via the **propositions-as-types** principle. This, as its name suggests, says that propositions and types (in the logical and programming senses respectively) are equivalent. In this section we go on to look at how the ideas above have more recently been brought together.

Basically, we use Heyting's definitions of the logical constants, so we have things such as (where $\mathcal{P}$ maps propositions to their proofs)

$$\begin{aligned}
\mathcal{P}(\mathcal{S} \wedge \mathcal{T}) &= \{(a,b) \mid a \in \mathcal{P}(\mathcal{S}) \text{ and } b \in \mathcal{P}(\mathcal{T})\} = \mathcal{P}(\mathcal{S}) \times \mathcal{P}(\mathcal{T}) \\
\mathcal{P}(\mathcal{S} \vee \mathcal{T}) &= \{i(a) \mid a \in \mathcal{P}(\mathcal{S})\} \cup \{j(b) \mid b \in \mathcal{P}(\mathcal{T})\} = \mathcal{P}(\mathcal{S}) + \mathcal{P}(\mathcal{T}) \\
\mathcal{P}(\mathcal{S} \Rightarrow \mathcal{T}) &= \{\lambda x\,.\,b \mid \text{if } a \in \mathcal{P}(\mathcal{S}) \text{ then } b(a) \in \mathcal{P}(\mathcal{T})\} = \mathcal{P}(\mathcal{S}) \to \mathcal{P}(\mathcal{T})
\end{aligned}$$

and so on. We see here that the witnesses to the truth of a proposition are made explicit and shown as terms, such as (a, b) in the proof of $\mathcal{S} \wedge \mathcal{T}$, which we write

$$(a, b) : \mathcal{S} \wedge \mathcal{T}$$

When we read this as a type we have

$$(a, b) : \mathcal{S} \times \mathcal{T}$$

just as in SML. So, for example, the third term in each of the lines above,

$\mathcal{P}(\mathcal{S}) \times \mathcal{P}(\mathcal{T})$, uses an interpretation that we would expect to see if we remember that we are looking for a link between propositions and types. The semantics introduces constructors such as $\times$, $+$, function space and so on.

The quantifiers, when thinking of propositions as types, are modelled by the operators $\prod$ and $\sum$, universal and existential respectively. These operators can be thought of as acting like quantifiers ranging over some type, that is instead of saying 'For all x, $\mathcal{T}(x)$' we say 'For all x in $\mathcal{S}$, $\mathcal{T}(x)$'. Extending the definitions above we get (and following Heyting again):

$$\begin{aligned}
&\mathcal{P}((\forall x \in \mathcal{S})\mathcal{T}(x)) = \{\lambda x \,.\, b(x) \mid \text{if } a \in \mathcal{P}(\mathcal{S}) \text{ then} \\
&\quad b(a) \in \mathcal{P}(\mathcal{T}(a))\} = \textstyle\prod(\mathcal{P}(\mathcal{S}), \mathcal{P}(\mathcal{T})) \\
&\mathcal{P}((\exists x \in \mathcal{S})\mathcal{T}(x)) = \{(a, b(a)) \mid a \in \mathcal{P}(\mathcal{S}) \text{ and} \\
&\quad b(a) \in \mathcal{P}(\mathcal{T}(a))\} = \textstyle\sum(\mathcal{P}(\mathcal{S}), \mathcal{P}(\mathcal{T}))
\end{aligned}$$

The final gloss we add is to interpret propositions not only as types but also as specifications. The reason that we can do this is that the type language is so rich, far richer than with SML for instance. As we shall see this means that we now have a single language in which to specify and construct programs.

9.5.5 The formal system

The formal system which defines the syntax and semantics of this powerful language is usually presented in a natural deduction style.

We should also say something about the expressions that appear in the language of the system. As usual expressions are built from a set of given constants, variables such as x and y and abstraction and are combined by the operation of application. If e is an expression which contains free occurrences of the variable x, then $(x)e$ is an expression in which the free occurrences of x in e are now bound. Application is then as usual, that is $(x)e(a)$ is an application of $(x)e$ to a, and it is computed by replacing all free occurrences of x in e by a (with the usual rules for avoiding variable capture).

For example, the usual natural deduction rules for introducing $\wedge$ might be

$$\frac{\mathcal{S} \qquad \mathcal{T}}{\mathcal{S} \wedge \mathcal{T}}$$

and in the formal system this would be written, making the proof witness explicit,

$$\frac{a : \mathcal{S} \qquad b : \mathcal{T}}{(a, b) : \mathcal{S} \wedge \mathcal{T}}$$

As another example we have the rules for $\rightarrow$

$$\frac{\begin{matrix}[\mathscr{S}] \\ \\ \mathscr{T}\end{matrix}}{\mathscr{S} \rightarrow \mathscr{T}}$$

which is written

$$\frac{b(a): \mathscr{T}[a: \mathscr{S}]}{\lambda(b): \mathscr{S} \rightarrow \mathscr{T}}$$

So, for example, we can construct a derivation as follows:

$$\frac{(x)x(x): \mathscr{S}[x: \mathscr{S}]}{\lambda((x)x): \mathscr{S} \rightarrow \mathscr{S}}$$

and we can see how this fits with the definition that Heyting gave to $\rightarrow$; if a is a proof witness of $\mathscr{S}$ then applying $\lambda((x)x)$ to a gives a back, which is a proof witness of $\mathscr{S}$, so $\lambda((x)x)$ is properly a proof witness of $\mathscr{S} \rightarrow \mathscr{S}$. We see here a simple example of a proof witness being a program, that is $\lambda((x)x)$ is the program for the identity function.

The inputs and outputs, so to speak, of the rules are known as **judgements**. So, the above $(x)x(x): \mathscr{S}[x: \mathscr{S}]$ and $\lambda((x)x): \mathscr{S} \rightarrow \mathscr{S}$ are judgements. The first, since it depends upon another judgement, is called a **hypothetical judgement**. There are four readings of the judgement form $a: \mathscr{S}$ which are:

(1) a is a proof witness of proposition $\mathscr{S}$

(2) a is an object of type $\mathscr{S}$

(3) a is an element of the set $\mathscr{S}$

(4) a is a program that satisfies $\mathscr{S}$ (as a specification)

The first two express the idea of propositions as types. The third can be seen as a reading of the first if we consider that, thought of as a set, the elements of a proposition are its proofs. The final reading is a more suggestive version of the second made possible because of the computational nature of proofs and the very expressive types.

Apart from logical rules such as these we also have the natural numbers as a basic type. If we think of the type N to be 'logically' the predicate 'there is a natural number' then we usually have

$$N(0) \qquad \frac{N(x)}{N(\mathrm{succ}(x))}$$

which we have in the formal system as

$$0:N \qquad \frac{x:N}{\mathrm{succ}(x):N}$$

We also have mathematical induction:

$$[P(0) \wedge (\forall k)(P(k) \rightarrow P(\mathrm{succ}(k)))] \rightarrow (\forall x)P(x)$$

which becomes

$$\frac{n:N \quad d:C(0) \quad e(a,b):C(\mathrm{succ}(a))[a:N, b:C(a)]}{\mathrm{rec}(n,d,e):C(n)}$$

Here $\mathrm{rec}(n, d, e)$ is an example of a **non-canonical** expression, that is an expression which can be evaluated or converted into a simpler (more canonical) form. In this case, as you would expect with 'rec' denoting primitive recursion, we have (with $\gg$ denoting 'converts to')

$$\mathrm{rec}(0, d, e) \gg d$$
$$\mathrm{rec}(\mathrm{succ}(a), d, e) \gg e(a, \mathrm{rec}(a, d, e))$$

For example, if we put $d \equiv 0$ and $e \equiv (x, y)\mathrm{succ}(y)$ then we have

$$\mathrm{rec}(0, d, e) \gg d \equiv 0$$

and

$$\begin{aligned}&\mathrm{rec}(\mathrm{succ}(0), d, e) \gg\\ &(x, y)\mathrm{succ}(y)(0, \mathrm{rec}(0, d, e)) \equiv\\ &\mathrm{succ}(\mathrm{rec}(0, d, e)) \gg\\ &\mathrm{succ}(d) \equiv\\ &\mathrm{succ}(0)\end{aligned}$$

In fact we can show that $(n, m)\mathrm{rec}(n, m, (x, y)\mathrm{succ}(y))$ behaves like addition.

9.5.6 Algorithm development – a sorting example

The formal system very briefly introduced above was developed by Martin-Löf (1985) and is called, variously, ITT, for intuitionistic type theory, or CST, for constructive set theory. Algorithm development in ITT consists of writing a specification (which will look like what may be viewed as a typed predicate calculus sentence) for a task and then, considering the specification as a type, demonstrating that the type is non-empty. If the demonstration is successful then an object of the appropriate type (equally a program that satisfies the specification, equally a proof that the statement of the, as we are reading it, typed predicate calculus sentence is true) will have been constructed.

In the course of proving the sentence it will probably, for all but the most trivial cases, be necessary to have many theorems about the objects that are mentioned. Some of these theorems will be about basic objects and will, hence, already be available since basic objects are defined by stating some of their properties (specifically, how to construct them and how to prove two objects equal).

For derived objects, that is objects which are not basic, the fact that they exist means that we have some sentences which describe some of their properties. Indeed, the activity of programming in ITT must be viewed as the construction of a hierarchy of objects, each with various theorems describing them, such that more complex derived objects are built from simpler ones.

So, when we start to develop an algorithm, we must assume that we are working in an environment where various properties of various objects are known, either by being basic properties or because, at some previous time, we have demonstrated that various derived objects have these properties.

The first task is to write down the specification for the program to be developed. In English we might express the specification of an algorithm to sort lists of natural numbers as

> 'write a program which, for any list x of natural numbers, gives a list l of natural numbers such that l is a permutation of x and l is ordered'

In ITT we would say that we must prove that the type

$$\prod(\text{List}(N), (x)\sum(\text{List}(N), (l)(\text{Perm}(x, l) \wedge \text{Ordered}(l)))) \qquad \textbf{(9.10)}$$

is not empty, and we do that by constructing an object of that type. This object will then be, under the alternative reading, a program which satisfies the specification. It turns out, as you may have guessed, that since this construction process is to be carried out entirely within this formal system it is a long and complicated process. There are many research level projects currently being carried out whose aim is to transfer more and more of the burden of these formal constructions onto programming systems so as to allow a programmer more freedom to think and provide the inspiration needed to make the constructions possible. For this reason, that is that the constructions are hard and also, in all their detail, unpleasant to look at, we shall only go as far as presenting the development of a type, or specification, for a sorting algorithm and the first simple steps in the construction of a program.

Before we can carry on with our construction, we must know what Perm and Ordered are. Ordered should be provable of a list x if x is nil or x is a singleton or its first two elements are in order and its tail is ordered, which we write (using $\equiv$ for definitional equality)

$$\text{Ordered} \equiv \\ (l)\text{listrec}(l, \text{True}, (x, y, z)\text{listrec}(y, \text{True}, (x', y', z')((x \leqslant x' =_B \text{true}) \wedge z)))$$

To see how the English and the definition match up first note that we have

$$\text{listrec}(\text{nil}, a, (x, y, z)c(x, y, z)) = a$$

and

$$\text{listrec}(b\,.\,l, a, (x, y, z)c(x, y, z)) = c(b, l, \text{listrec}(l, a, (x, y, z)c(x, y, z)))$$

where . is infix cons.

The type True has only one element, that is

$$\text{True} \equiv \{\text{it}\}$$

and is trivially non-empty since the property that defines it is

$$\text{it: True}$$

that is the sentence True is trivially always provable. Thus, in the definition of Ordered above we are stating that Ordered(nil) is simply the type True and thus

$$\text{it: Ordered(nil)}$$

so Ordered(nil) is true.

By repeatedly applying the listrec we have

$$\text{Ordered(nil)} = \text{True}$$
$$\text{Ordered}(a\,.\,\text{nil}) = \text{True}$$
$$\text{Ordered}(a\,.\,(b\,.\,c)) = (a \leqslant b) =_B \text{true} \wedge \text{Ordered}(b\,.\,c)$$

where the type B (for boolean) is another enumerated type such that

$$B \equiv \{\text{false}, \text{true}\}$$

Similarly, we define Perm by

$$\text{Perm} \equiv (a, b)(\text{listrec}(a, \text{True}, (x, y, z)(\text{occurrencesof}(x, a) =_N \\ \text{occurrencesof}(x, b) \wedge z) \wedge \text{len}(a) =_N \text{len}(b))$$

As with our previous natural deduction derivations from Chapter 5, a good, overall strategy is to work in a goal-directed fashion, bottom up. So, given that the type 9.10 needs to appear at the root of a derivation we need to find a rule which has a judgement of the form

$$e: \prod(A, B)$$

as its conclusion. Such a rule is called $\prod$-intro and has the form

$$\frac{b(x): B(x)[x: A]}{\lambda((x)b): \prod(A, B)}$$

so our first step is to write the root of the derivation as, where b is to be discovered as the derivation proceeds,

$$\frac{b(x): \sum(\text{List}(N), (l)(\text{Perm}(x, l) \wedge \text{Ordered}(l)))[x: \text{List}(N)]}{\lambda((x)b): \prod(\text{List}(N), (x)\sum(\text{List}(N), (l)(\text{Perm}(x, l) \wedge \text{Ordered}(l))))}$$

Now we have reached the stage where the construction starts to get very detailed and difficult and the reader should note that this section forms the basis of the more detailed introduction in Reeves (1989). As we said before, we have to leave the development here and allow the interested reader to follow the references given earlier on their own.

SUMMARY

- We have seen four methods of deciding validity in predicate calculus, namely axiom systems, semantic tableaux, natural deduction (or sequent calculus) and resolution. All these methods are logically equivalent but they differ substantially when viewed as algorithms for practical implementation. 'State of the art' at the time of writing is the *connection method* which can be viewed as an efficient development of the tableau method that avoids explicit and wasteful duplication of subformulas and branches.
- The basic reason why fully automatic theorem-proving is currently too slow to be practical is that it is difficult, in a fully automatic uniform procedure that does not know the meaning of the formulas it is manipulating, to control the size of the search space. Currently, the most successful systems for using logic to reason about programs are *interactive*. Both partners do the part they are good at. The machine does the book-keeping and checks that the rules are correctly applied;

the person interacting with it chooses what to do next on the basis of understanding and experience of the domain. The LCF system described in this chapter goes further than this. Based on natural deduction, it not only does the basic book-keeping when rules are applied but also allows general-purpose proof techniques, such as working backwards, to be used, and simple strategies to be combined together into more complex ones.

- One important application of modal logic is in *temporal reasoning*. For example, in reasoning about the execution of concurrent programs, the possible worlds are different states of a computation at different points in time and the accessibility relation represents the ordering of time points. The modalities are used to *tense* statements, for example $F\mathscr{S}$ is true if $\mathscr{S}$ is true at some future time.
- Another application of the ideas of modal logic in computer science is *dynamic logic*. Here the possible worlds are different states of a computation and the accessibility relation represents the program which determines, for each state, the state that succeeds it. The $\Box$ of modal logic now depends on the program α and is written $[\alpha]$. $[\alpha]\mathscr{S}$ is true in state s if $\mathscr{S}$ is true in the state that results when α is executed in state s. Dynamic logic is expressive enough to represent many programming constructions such as 'if . . . then . . . else'.
- Intuitionistic logic arose out of the constructivist movement in mathematics which rejected non-constructive existence proofs or those based on the law of the excluded middle. Recently, a connection between intuitionistic mathematics and programming has come out of the observation that propositions and types (in the programming sense) are equivalent. Algorithm development in this formal system, which is based on natural deduction, consists of writing a specification in logical notation and then, considering this as a type, proving that it is non-empty. Because the underlying logic is constructive, the proof, if it can be carried out, must embody the explicit construction of an object of the appropriate type.

APPENDIX A

Introductions to Standard ML and Prolog

A.1 Introduction

In this appendix we shall describe and present programs in Standard ML and Prolog for the various algorithms introduced in the body of the book and for those asked for in answers to exercises.

We begin with a brief run through the basic features of SML and Prolog, so that the reader who is fairly familiar with these languages may understand the programs that we present. For those of our readers who are not familiar with either of these languages we would recommend the texts by Wikström (1987) and Bratko (1990) as excellent introductions to SML and Prolog respectively.

A.2 Standard ML

A.2.1 Simple values

A **program** in Standard ML (hereafter SML) is a sequence of **declarations** which are always read from top to bottom and left to right. Each declaration refers to and updates a **value environment**, which is a function which maps names to values. The simplest form that such a declaration can take is to associate a name with a basic value, such as an integer:

```
val n = 2;
```

This declaration associates the name n with the value of 2, the association being recorded in the environment. When this declaration is presented to the SML implementation it not only updates the environment but its **type** is also calculated, printed and added to a **type environment**, which maps names to types. So, having checked that the above declaration is well formed, that is conforms to the grammar of SML, the implementation prints some type information; in this case we have

```
val n = 2 : int
```

which tells us that n is associated with the value of 2 in the value environment and the type of n is int, that is n is associated with the type 'int' in the type environment.

There is a particular form of declaration which looks as though it should be called something such as 'expression'. For example

```
2;
```

is of this form. In fact any input of this form is treated by SML as a declaration of the hidden name it. This name is a standard one that exists in every environment and any input such as the one above is taken to be an abbreviation of the declaration

```
val it = 2;
```

or more generally

```
val it=EXP
```

where EXP is any SML expression.

A.2.2 Function values

As well as simple values of the types of int, string, bool and so on we have values of function type. For instance, we can define a successor function s by

```
fun s(x) = x+1;
```

Again, the environment is updated so that the name s is associated with a value; in this case the value is the function which adds one to its single argument. Given the above declaration the interpreter will return

```
val s = fn : int -> int
```

that is that s is a function of type integer to integer, as we would expect. Note that instead of writing out the value of s the interpreter writes just fn.

The language SML is known as a **strict** language. This means that the arguments to any function are evaluated before the body of the function is evaluated. This means, for instance, that if one of the arguments is undefined then the whole function is undefined *whether or not the argument is used in the body of the function.* To see the difference consider the function s again and imagine applying it to the expression 1 div 0. As we would hope, the evaluation of the expression s(1 div 0) fails, but this would have happened even in a language that was not strict. However, now consider applying the function defined by

```
fun t x=2;
```

to 1 div 0. In SML the expression t(1 div 0) is also undefined, even though the argument is unused in the body of t, proving that SML is strict. In a non-strict language, where the arguments to a function are evaluated only if they are needed in its body, the expression would have evaluated to 2.

While we are talking about the notion of an expression being undefined we should say something about how this is handled in SML. If you try to evaluate 1 div 0 then you will get the message

```
Failure : div
```

from the interpreter, which means, as we expected, that the evaluation of the function div for this pair of arguments has failed. In the jargon of SML we say that an **exception** has been **raised**. As a first approximation to how this system works we can assume that if we are evaluating an expression and an exception is raised then the evaluation fails and the exception is passed to the enclosing expression. If the enclosing expression cannot do anything with the exception then it is passed to the next enclosing expression and so on. Eventually the exception will reach the outer level and at this point the interpreter forms the message that we have just seen above.

In SML we do not have to rely just on exceptions from system-defined functions; the user can define exceptions too. For instance we may want to fail in the evaluation of a function if the value that is given as its argument is, in some way, too big. We would first declare the exception, say large, and then raise it at the appropriate point in the function

```
exception large;
fun test n=if n>10000 then raise large else n div 5;
```

which would give values such as

```
test 6;
1 : int
test 12000;
Failure: large
```

In fact exceptions can be used in far more sensitive ways than the simple one shown here, but this is something we must leave the reader to learn from the recommended books.

This simple example of the definition of s also shows an important feature of SML, namely that of **type inference**. This is the mechanism by which the interpreter, in the example above, was able to calculate the type of the function s. The reasoning was as follows: assume the type of x is unconstrained. Then, we see that x is involved in an addition and one of the values in the addition is 1 which has type int. Therefore, for the expression to type correctly the other value in the addition must have the type int too. Since int is one of the types that x can have, because its type is unconstrained, we can consistently type the whole expression to give the answer as above.

These definitions using fun can also be recursive, so that the declaration

```
fun t x= if x>0 then t (x-1) else 1;
```

would result in the environment being updated so that t was associated with the function of one argument that always has value one.

SML has a way of writing such a function which can be thought of as SML's way of writing λ-expressions which some of our readers may have come across. Where we might write $\lambda x . \lambda y . x + y$ ordinarily we can write

```
fn x=> fn y=> x+y
```

in SML. So, if we write the declaration

```
val f = fn x => fn y => x+y;
```

then f has the value denoted by $\lambda x . \lambda y . x + y$ and so is, of course, equivalent in meaning to

```
fun f x y=x+y;
```

and our function t above can also be defined by

```
val t=fn x=>1;
```

Functions are treated as 'first-class citizens' in SML. That is, they, like

$$\frac{f{:}\,(\text{'a} \to \text{'b}) \qquad x{:}\,\text{'a}}{fx{:}\,\text{'b}}\,(1) \qquad \frac{x{:}\,\text{'a} \qquad e{:}\,\text{'b}}{\text{fn } x \Rightarrow e{:}\,(\text{'a} \to \text{'b})}\,(2)$$

Figure A.1.

any other type of value, can be taken as argument and given as result by functions. So, it makes perfect sense to have the declarations

```
fun s x=x+1;
fun twice f y=f(f(y));
```

The interpreter will write

```
val s=fn : int-> int
val twice=fn : ('a->'a)->'a->'a
```

which means that the environment is updated so that s refers to the successor function and twice refers to a function which takes a unary function of any type 'a (where 'a can be thought of as a type variable, that is an expression that can stand for any type) and a value of type 'a and returns the result, also of type 'a, of applying the function twice to the argument in the obvious way. This example also introduces the idea of **polymorphic types**, that is types which contain variables and, hence, can themselves stand for a range of types. So, since s has type int-> int, it can properly have twice applied to it, because the type of s and the type of the first argument of twice can be unified, that is there is a substitution of the type variable 'a that makes it the same as the type of s. The result of making the declaration

```
twice s;
```

is to associate it with the function which when applied to x has s(s(x)) as value, that is the same value as the value of fn x=> s(s(x)), and to write

```
fn : int-> int
```

that is to give the type of twice s.

As something of a digression we shall present some rules for finding such types. These rules will be in a natural deduction style but with objects and their types in the place of propositions.

If we write f : t to mean the object f has type t then we have two rules as shown in Figure A.1. Now, we can use these rules to work out the value of twice in the following way. First, we remember that twice can also be defined by

```
val twice = fn f => fn y => f(f(y));
```

```
                                      f:('f->'e)     y:'f
                                      ------------------- (1)
                       f:('e->'d      f(y):'e
                       ---------------------- (1)
              y:'c     f(f(y)):'d
              ------------------- (2)
f:'a          fn y=>f(f(y)):'b
------------------------------- (2) where 'b=('c->'d)
fn f=>fn y=>f(f(y)):('a->'b)
```

Figure A.2.

First, we do a piece of 'rough working' where we leave types as unconstrained as possible but where the rules (1) and (2) are obeyed. Here 'a, 'b, 'c, 'd, 'e and 'f are type variables and the working is shown in Figure A.2.

To make this a proper derivation we need to make sure that we have the same type for all occurrences of f and all occurrences of y. For this to happen we need 'a to match with 'e->'d and with 'f->'e.

This means that we have 'd='e='f. Also we need 'c to match with 'f, so 'c='f. This in turn means that 'a='e->'e and 'b='c->'d='e->'e. Then Figure A.3 shows a proper derivation of the type of twice which, up to a consistent renaming of type variables, is as we expected.

This use of polymorphic types does carry some overhead in the sense that the programmer has to be careful to disambiguate contexts when necessary otherwise type inference may not succeed. For instance if we made the declaration

```
fun g x = x * x;
```

then, since * is overloaded, this definition is ambiguous, for we cannot tell whether the operation to be carried out is, say, multiplication of reals or of ints, which are, of course, very different operations. (To see this consider the fact that in real multiplication there are extra rules about where the decimal point goes which integer multiplication does not have.) The interpreter will respond to this declaration with a message saying that it cannot decide on the type of

```
                                      f:('e->'e)     y:'e
                                      ------------------- (1)
                     f:('e->'e)      f(y):'e
                     ---------------------- (1)
             y:'e     f(f(y)):'e
             ------------------- (2)
f:('e->'e)       fn y=>f(f(y)):('e->'e)
------------------------------------------- (2)
fn f=>fn y=>f(f(y)):(('e->'e)->'e->'e)
```

Figure A.3.

the * and the declaration will fail. This problem can be overcome by the use of type tags which convey type information to the interpreter. For instance, we can disambiguate the attempted declaration above by writing

```
fun g (x : real) = x * x;
```

to get the response

```
fun g = fn : real -> real
```

as expected. We could have got the declaration to work by tagging the result type of the function, rather than its argument type, by writing

```
fun (g x) : real = x * x;
```

or we could also have written

```
fun g x = (x : real) * x;
```

It is sometimes more usual to think of certain functions as being **infix**, that is written between their arguments. We can arrange for this in SML by using the infix directive, which tells the interpreter that the function named is going to be used in an infix fashion. We would write

```
infix plus;
fun x plus y : int = x + y;
2 plus 3;
```

to get the required effect.

A.2.3 Datatypes

The next idea in SML that the reader needs to understand, to allow them to have some idea of the meaning of the SML programs given in this book, is that of **datatype declarations**.

We can think of a datatype declaration as describing a way of making a new type from a mixture of old types and, since they can be recursive, from the new type itself. For instance, if a term could be either a string or a term applied to a term we would write

```
datatype term = var of string | app of term * term;
```

and this would introduce two new values, called **constructors**, named var and app above, which take a string and a pair of terms, respectively, and give a term. So, if we have the string "abc" then it can be made into a term using the constructor var to give var "abc". Then, having the term var "abc" we can make another term using app to give app(var "abc", var "abc").

In the parser for propositional calculus, which we present completely later, we have the following datatype, which you should now be able to understand:

```
datatype SENT = Prop of string |
                Not of SENT |
                And of (SENT * SENT) |
                Or of (SENT * SENT) |
                Imp of (SENT * SENT) |
                Eq of (SENT * SENT);
```

Having constructors to form new types not only makes it easier uniformly to build up quite complicated types, but it allows the introduction of a technique which makes it easier to define functions which work on such types using the idea of **matching**. Imagine that we have the datatype term as introduced above and, for some reason, we wish to recover the components that go to make up the values of type term. We can do this by defining a function which systematically works through all the possible forms of values of type term and describes an operation for each such form. So, since we have two possible forms of values of type term, namely var s and app(t1,t2), s is a string and t1 and t2 are of type term, we have two clauses in the function definition:

```
fun parts (var s) = [s]
  | parts (app(t1,t2)) = [parts t1,parts t2];
```

The first clause tells us that if the argument has the form var s then its parts are just the string s which it returns in a list. However, if the argument is in the form of an application involving t1 and t2 then the list of parts returned is calculated by finding, in turn, the parts of t1 and t2. This form of definition at once makes it clear what to do with each form of term and allows us to see that we have looked at all possibilities, since we have two parts to the definition of term and two clauses in the function parts. To complete this idea we also need to know that a variable matches anything and a constant matches only itself.

This technique is used heavily in the programs in this book since we are mainly taking sentences or formulas or terms in logical languages and manipulating them, which means breaking them into their component parts and rebuilding these parts into other terms. The function decide given in Chapter 2 is probably the simplest example of this sort of operation.

A.2.4 Other forms of expressions

As well as the forms of expression and declaration that we have introduced so far there are others, most of which you probably expected to see and whose meaning is likely to be obvious to you. Probably the most common expression used is the if–then–else form. So, we would use it in definitions such as

```
fun fact n = if n = 0 then 1 else n * (fact (n - 1));
```

Of course in this case we could have used the matching mechanism to define the same function as in

```
fun fact 0 = 1
  | fact n = n * (fact(n - 1));
```

The choice between these two forms is largely one of style, but the latter can be useful with more complicated types to ensure that all possibilities of argument form have been taken into account, as we said before.

There is a restriction on the typing of the conditional expression, and that is that both 'arms' of the conditional must have the same type. So the expression

```
if true then 1 else 2.0;
```

would not be allowed by the interpreter because the 'true' arm gives a value of type int and the other arm a value of type real. Since these two types are mutually exclusive the expression has no type and is rejected.

We said above that SML is strict, but the conditional expressions and boolean expressions are an exception to this. So, when evaluating

```
if true then 1.0 else (1.0/0.0);
```

the interpreter gives the value 1.0 rather than an error, as it would if the expression were thought of as the 'function' if–then–else being applied to the arguments true, 1.0 and (1.0/0.0). So, the rule is that the boolean test is first evaluated and then the appropriate arm, the other arm being discarded. Also, boolean expressions are treated in a similar way so we should consider, for instance, the definition of the function andalso to be

```
infix andalso;
fun x andalso y = if x then y else false;
```

so the value of

```
false andalso ((1.0/0.0) = 0.0);
```

is false rather than undefined.

Remember that declarations are read from top to bottom and a name must be defined before it is used. This means that a definition that is **mutually recursive**, that is one in which, say, two functions are defined, each referring to the other, does not seem to be allowable. This is overcome by the use of the and construct for combining declarations. So, whereas

```
fun f x=if x=0 then 1 else g(x-1);
fun g x=if x=0 then 0 else f(x-1);
```

would lead to a message from the interpreter to the effect that g is undefined in the definition of f, we can make the declaration work by writing

```
fun f x=if x=0 then 1 else g(x-1)
and g x=if x=0 then 0 else f(x-1);
```

We use this construct a great deal in the programs in this book, especially for the parser, which contains many definitions that are mutually recursive.

A further addition to forms of expression is provided by the 'let–in–end' form. This allows definitions to be made locally, which enhances the modularity of programs and can make them much more efficient. For instance, we might write

```
fun f y x z=if y=2 * (x div z) then 3 * (x div z) else 0;
```

but the same effect, as far as the definition of f is concerned, would be produced by writing

```
fun f y x z=let val u=x div z in
                 if y=2 * u
                 then 3 * u
                 else 0
             end;
```

Here, u is not defined either before or after the declaration of f and so reference to it is a mistake. The advantage is that x div z is computed only once, whereas originally it had to be computed twice. Clearly, with very much more complicated repeated computations, the use of this mechanism can save a lot of computing effort. Another reason for using this construct is for enhancing modularity, that is the property of programs that makes them separable into independent parts. This means that different programmers can work on the parts of a program without fear of creating mistakes or ambiguities. For instance, a programmer working on a part of a program might have used the name x to refer to some value and another programmer working on a different part might have used x to refer to something different, and the effect of combining these two parts might not be what was intended. So, this construct can be used to hide names that are purely local to particular declarations, especially when there is no need for them to be available anywhere else.

We can also use local declarations in other declarations by using the 'local–in–end' form, as in

```
local fun (g x) : int = x * x in
      fun f a b c = ((g a) * (g b) * (g c) div 2)
end;
```

or in

```
local val e = 2 div 3 in val y = e * e end;
```

which again can enhance modularity (and readability) and also improve efficiency.

A.2.5 Lists

Lists form a predefined type in SML. Lists are formed by using the constructors nil, which is the empty list, and ::, which is a function from elements and lists to lists, also known as 'cons'. Both these constructors are polymorphic, with the proviso that all the elements of a list must have the same type. For instance we can form lists of ints such as

```
1::(2::(3::nil));
```

or of strings such as

```
"a"::("b"::("c"::("d"::nil)));
```

or of lists of bools such as

```
(true::nil)::((false::nil)::nil);
```

and these expressions would be typed as int list, string list and (bool list) list respectively. If you try evaluating these expressions using an SML interpreter you will find, however, that the value printed is not in the same form as given above. In fact, the responses will show the lists as

```
[1,2,3]
["a","b","c"]
[[true],[false]]
```

and this is because there is a shorthand way of writing lists which means that nil becomes [ ] and e::nil becomes [e] etcetera.

We can define functions over lists either by using the matching mechanism or by using the 'if–then–else' expression. For instance, we can define the function which gives the length of a list over any type as

```
fun length(a::l) = if l = nil then 1 else 1 + (length l);
```

or as

```
fun length nil = 0
  | length (a::l) = 1 + (length l);
```

or as

```
fun length (a::nil) = 1
  | length (a::l) = 1 + (length l);
```

and each of these functions will have type

```
('a list) -> int
```

The question of which of these definitions is 'best' is hard to answer but the second will pass the interpreter without comment while the first and third will provoke a message such as

```
*** Warning: Patterns in Match not exhaustive
```

which means that there is the possibility that the writer has not given a value of the function for all possible forms of the argument. The reason that the interpreter does not give this warning for the second declaration is that we should imagine the type 'a list as defined by

```
infix ::;
datatype 'a list = nil | op :: of ('a * 'a list);
```

So as far as the interpreter is concerned there are two forms which define lists: either nil alone or an expression such as e1::e2. As you can see the first and third declarations for length do not mention both of these forms and so the interpreter gives a warning just in case you did not mean to leave out some possible forms. In fact we can see that in the first and third examples the functions are not defined for lists which are nil, which may be a mistake.

Two functions that are very commonly applied to lists are map and append, denoted by @ in SML. The idea of map is to encapsulate the operation of 'doing the same thing to all elements of a list'. For instance, we may have a list of ints and want to square them all. A function to do this, called squarelist, might be

```
fun squarelist nil = nil
  | squarelist ((n : int) :: l) = (n * n) :: (squarelist l);
```

We can make this declaration look a bit more general by taking the squaring part and using a local declaration for it:

```
local fun square (n : int) = n * n in
  fun squarelist nil = nil
    | squarelist (n::l) = (square n) :: (squarelist l)
end;
```

Now consider a function that takes a list of booleans and negates each of them:

```
local fun negate b = if b then false else true in
   fun negatelist nil = nil
     | negatelist (b::l) = (negate b) :: (negatelist l)
end;
```

If you compare these two definitions a pattern becomes clear. They are both of the form

```
local fun f x =< some expression > in
  fun g nil = nil
    | g (x::y) = (f x) :: (g y)
end;
```

This is still not as general a form as is possible though. The auxiliary function f above is only used in the declaration of g, but if we make the auxiliary function an argument of g then g is more generally usable. So, we now have

```
fun g f nil = nil
  | g f (x::y) = (f x) :: (g y);
```

and the two functions we had before can be defined by

```
fun squarelist l = g (fn (n : int) => n * n) l;
fun negatelist l = g (fn b => if b then false else true) l;
```

Since this function that we have called g turns out to be so useful it is already defined within the SML interpreter and is the function called map. The other function is @ which can be thought of as defined by

```
infix @;
fun nil @ m = m
  | (a::l) @ m = a::(l @ m);
```

and simply takes two lists, of the same base type, and joins the left one to the right one keeping the same order between the elements.

Armed with these basic functions you should now be able to understand the functions over lists used in the programs in this book, with more or less work. You will see that lists form a very adaptable type which can be used to imitate many different structures and, historically, they were used to imitate tree structures. For instance, a binary tree might be imitated by a list of lists, each list denoting a path from the root through the tree. However, in SML we can use the datatype mechanism to represent trees much more directly, and since, via this method, we also have all the typechecking mechanism working for us we can have more confidence in our solutions.

A simple binary tree might be defined by

```
datatype bintree = null | node of (int * bintree * bintree);
```

which means that examples of values of type bintree might be

```
null
node(1,null,null)
node(1,node(2,null,null),node(3,null,null))
```

which might be displayed in two dimensions as

and so on for more complicated shapes of tree. Now we can write functions that manipulate the contents of nodes rather as we did with map, the function that manipulated the contents of lists, without changing their structure or shape. (This sort of function, a homomorphism, turns out to be a central notion throughout computer science and logic.) A function to square the contents of each node might be squarenodes given by

```
fun squarenodes null = null
  | squarenodes (node(n,t1,t2)) = node(n * n, squarenodes t1, squarenodes t2);
```

and again we can see how the definition of the function follows exactly the structure of the tree. However, just as we did with lists, we can write a more abstract function which does an arbitrary operation, provided as an argument, to all nodes of a tree while preserving the tree structure. We call this maptree and it is defined by

```
fun maptree f null = null
  | maptree f (node(n,t1,t2)) = node(f n,maptree f t1, maptree f t2);
```

and using this our first function can be defined by

```
fun squarenodes t = maptree (fn n => n * n) t;
```

The other main activity that takes place when we traverse trees is to build new datatypes from their parts. For instance, we may wish to construct a list which contains all the values in the nodes of a value of type bintree. Again we need to write a function which takes into account all the possible forms of tree but this time instead of preserving the structure of the tree it produces values from the list datatype. A reasonable definition might be

```
fun getnodes null = [ ]
  | getnodes (node(n,t1,t2)) = n::((getnodes t1)@(getnodes t2));
```

and when we apply it to each of the three trees given above we get the values [ ], [1] and [1,2,3] respectively.

As a final level of abstraction we can introduce a function which can be specialized to either maptree or getnodes, even though maptree preserves the structure of the tree and getnodes constructs something almost completely new. We can do this first by exploiting the structure of the bintree and also by using, quite heavily, the facility with which SML deals with functional parameters. We construct a function which takes two arguments, one which describes the value of the function at values of the form null and the other which describes a value for the form node(n, t1, t2). If these arguments are represented by f and s below then we know that the most information that can be passed to s are the three pieces of information that go to make up a node, namely the int value n and the two subtrees, values of type bintree. Since the information for a node which is null is fixed the f is a constant. Our definition for this most general function, which we name bintreerec, for bintree recursion, is

```
fun bintreerec f s null = f
  | bintreerec f s (node(n,t1,t2)) = s n (bintreerec f s t1) (bintreerec f s t2);
```

and using this we can define both maptree and getnodes by abstracting the appropriate auxiliary function from their definitions which gives us

```
fun maptree g l =
    let val f = null
        fun s n t1 t2 = node(g n,t1,t2)
    in bintreerec f s l end;
```

and

```
fun getnodes l =
    let val f = [ ]
        fun s n t1 t2 = n::(t1@t2)
    in bintreerec f s l end;
```

The use of bintreerec gives us even more checks on the correctness of our definitions of functions over bintrees because, once bintreerec is properly defined, we have much more structure, which the typechecker can use, against which to judge the completeness of our subsequent definitions. This level of abstraction means that a few highly abstract definitions can be given and then specialized as needed to the more basic functions over a given datatype and this means that we, as programmers, have less to remember to check because the typechecker does much of the work for us. This, in turn, means that our programs are much less likely to be wrong and this is the great strength of languages such as SML.

We have introduced some of the main features of SML, but much more remains to be said and the textbooks referred to should be consulted for further details.

A.3 Prolog

This brief introduction is not intended as a stand-alone course on Prolog programming. The aim is to give sufficient explanation of the language, and relevant programming techniques, for the programs in the text to be used and understood by a reader with some experience of a procedural or functional language such as Pascal, C, Lisp or Miranda. Good books that give a more complete explanation of Prolog and cover a much wider variety of programming tasks are Bratko (1990), Sterling and Shapiro (1986) and the original Clocksin and Mellish (1981).

Prolog is a practical programming language whose theoretical basis lies in the notion of logic programming, an introduction to which is given in Chapter 7. Here we are concerned with the pragmatics of using Prolog to carry out calculations in which the objects manipulated are logical formulas. This section will not depend on your having read Chapter 7, but use of some basic logical terminology, as explained in Chapters 2 and 3, is unavoidable.

A Prolog program is a sequence of **clauses**, which play much the same role as function definitions in a functional language such as Lisp, Miranda and SML, or procedure definitions in Pascal and C. Clauses have a **head** and a **body** and are always terminated by a full stop. The head has the form of a logical atom, that is it consists of a predicate name followed by zero or more arguments which have the form of logical terms (see Chapter 3 and below). The body is a sequence of zero or more atomic formulas separated by commas and terminated by a full stop. The head and body are separated by the compound symbol ':-' which is omitted if the body is null. The following is an example of a complete Prolog program.

```
p(X) :- q(X,Y),r(Y).
```

```
q(a,b).
q(a,c).
r(c).
```

Variables in Prolog are denoted by strings beginning with a capital letter or the underline symbol '_'. All other strings not starting with a capital letter or the '_' symbol are constants in various categories such as predicate names, data, operators, numbers. Prolog programs can be read in both a declarative (that is, logical) and a procedural (that is, operational) way. The declarative reading is ultimately more productive, but the operational reading can be helpful provided that one remembers that Prolog 'procedures' are **satisfied** (by appropriate substitutions for their bound variables) rather than executed. So the first line of the program above can be compared with the definition of a function or procedure **p** with formal parameter **X**, the **q(X,Y),r(Y)** on the right of the ':-' symbol being analogous to the statements in the body of the function or procedure. The predicates **q** and **r** are like procedures with no body and so are analogous to primitive instructions. Many Prolog texts call the first line, that defines **p(X)**, a **rule** and the remaining clauses **facts** because the latter are unconditionally satisfied.

Declaratively, the first line of the program is read as 'For all terms that can be substituted for **X**, **p(X)** is satisfied if there is a term that can be substituted for **Y** such that both **q(X,Y)** and **r(Y)** can be satisfied'. The second line of the program says that **q(a,b)** is unconditionally satisfied. It can be seen that **p(a)** can be satisfied because there is a substitution for **Y**, namely **c**, that enables both **q(a,Y)** and **r(Y)** to be simultaneously satisfied. Although substituting **b** for **Y** satisfies **q(a,Y)** it does not satisfy **r(Y)** because there is no clause **r(b)** in the program. It should be noted that if, within a clause, a substitution is made for a particular instance of a variable symbol then the substitution is made throughout the clause. So, in the first line of the program above, the two occurrences of the variable symbol **X** either have the same instantiation or are both uninstantiated, and the same applies to **Y**. However, a particular variable symbol denotes different variables in different clauses, in a similar way, for example, to the local variables or formal parameters of a procedure in Pascal. There are no global variables in Prolog, although we see later that operator definitions can be given global scope.

A Prolog program is used by typing a **goal** in response to the interpreter's prompt. Following the procedural analogy this is the Prolog equivalent of a procedure call. An example is

```
?-p(a).
```

The compound symbol **?-** is the system's prompt and the **p(a).** is input by the user and is a request for the system to check that the formula **p(a)** follows logically from the program clauses. The reply, that **p(a)** does follow logically from the program, is output simply by the system as '**yes**'.

Operationally speaking the goal is **matched** against the heads of all program clauses starting at the top of the program until a match is found, in this case at the first line. The head of the first clause is **p(X)** and **X** is a variable so a substitution can be made for it to enable the match to succeed, in this case by substituting **a** for **X**. Then a similar matching process is carried out on the body of the clause, which is now **q(a,Y),r(Y)**, working from left to right. Starting again from the top of the program **q(a, Y)** is matched against the clause heads. It cannot match **p(X)** because the predicate symbol **q** is not the same as **p**. However, it matches **q(a,b)** provided that **Y** is instantiated to **b**. We now attempt to satisfy **r(b)** starting again at the top of the program. This time no match can be found for **r(b)** so we go back to **q(X, Y)** and see whether this could be satisfied in a different way. This going back to try other possibilities for goals to be satisfied is called **backtracking**, and is one of the ways in which Prolog differs radically from other languages, including SML. The matching process for **q(a, Y)** resumes from the point it previously got to at **q(a, b)**. Another match is found, this time with **Y** instantiated to **c**, so now we try to satisfy **r(c)** starting again at the top of the program. This time the search succeeds so the original goal **p(a)** succeeds and the system outputs **yes**.

The possibilities become more interesting when goals contain uninstantiated variables. The goal

```
?-p(Answer).
```

can be read as a request for a binding for the variable **Answer** that enables the formula **p(Answer)** to be satisfied. The resulting binding (the output from the computation) is printed by the system as **Answer=a**. The variable name **Answer** has no particular significance, of course. We use it to emphasize that one is not restricted to the $X, Y, Z, \ldots$ of mathematical or logical notation. Any other string starting with a capital or '_' could have been used.

Operationally the computation is very similar to that described above for **?-p(a)**. The goal **p(Answer)** is matched against the heads of all program clauses starting at the top of the program until a match is found, at the first line in this case. Uninstantiated variables always match and become, in effect, two aliases for the same variable. The body of the matching clause is now **q(Answer,Y),r(Y)** and an attempt is now made to satisfy this, working from left to right. Starting again from the top of the program **q(Answer,Y)** is matched against the clause heads. As before it matches **q(a,b)**, this time with **Answer** instantiated to **a** and **Y** to **b**. But **r(b)** fails and backtracking occurs in the same way as before. Eventually the original goal **p(Answer)** succeeds and the instantiation that enabled it to succeed, namely **Answer=a**, is output.

Chapter 7 will show that a Prolog program (or at least most of it) is a formula in a subset of first-order logic, and the interpreter that executes it is a simple-minded theorem prover. This means that Prolog programs can be understood and reasoned about at a logical level, as well as in the operational style exemplified above. The declarative (logical) reading of Prolog programs is the best one for constructing programs and convincing oneself of their correctness but, because the Prolog interpreter is not a complete theorem prover for first-order logic, the operational understanding is also necessary for reasoning about completeness and efficiency. The perfect logic programming system has not yet been invented.

In comparison with languages such as Pascal or SML there is no equivalent notion of type in Prolog. The only data structure is the **term**. A term is a constant, a variable, or a function symbol followed in parentheses by one or more terms. The definition is as it is in Chapter 3 or, less formally, like the notion of expression in mathematics. Prolog is essentially a system for doing calculations in a term algebra. There are several examples below that use terms but, as a simple example here, we can slightly elaborate the program above by replacing the clause **q(a,c)** by **q(f(a),c)**. The output that would then result from satisfying the goal

```
?-p(Answer).
```

would be **Answer = f(a)**. Note that no type declarations are required to say for example that **f** is a function symbol rather than a predicate symbol. The grammatical category of any given symbol in the program is inferred from its position in the program text. In this case **f** is a function symbol because a term is expected after the symbols **q(**. This is not the same as in logic where you have to say in advance what the function and predicate symbols are.

A.3.1 System predicates and unification

All Prolog systems make available to the programmer a predefined set of predicates and one has to think of the user-defined program as being an extension of these. One of the most important is =, which can be thought of as being internally defined in every program by the single clause

```
X = X.
```

Thus with no user-defined program at all, you can input the goal

```
?- a = a.
```

and get the output **yes** (because it matches **X = X.** with **X** as **a**) or

```
?- a = b.
```

and get the answer **no** because **a = b.** cannot match **X = X.** The goal

```
?- Y = a.
```

results in the output **Y=a** because the goal **Y=a** is matched against **X=X** and succeeds with **Y** instantiated to **X** and **X** to **a**. A more interesting example is

```
?- f(g(Y),h(b))=f(g(a),Z).
```

which gives the output

```
Y=a
Z=h(b)
```

because these are the variable instantiations required to make the left- and right-hand sides the same, that is to make **X=X.** succeed. This computation of the set of substitutions required to make two terms identical is known as **unification** and is discussed in more detail in Chapter 7. A goal matches a program clause in Prolog if the predicate names are identical and the arguments can be unified pairwise. Note that unification is more general than the matching in functional languages such as SML because in Prolog variables can occur in both sides to be matched. One limitation, though, is that function or predicate symbols in Prolog cannot be variables. Variables stand for terms, not for symbols (remember that a constant is a term).

In this book we are using Prolog to do calculations in a term algebra, with expressions formed from constants, variables and function symbols, so we will now concentrate entirely on this type of application and briefly cover some of the necessary techniques. In particular we want to be able to represent logical formulas and terms, and to take them to pieces in order to examine and operate on their component parts. We also have to carry out calculations with sets and sequences of terms.

A.3.2 Sets, sequences and lists

It is convenient to represent both sets and sequences as **lists**. Lists have to be represented in Prolog as terms since that is the only data type but, since manipulating lists is such a common requirement, all Prolog systems provide a special notation to make list programming easier and programs more readable. By convention a special constant [ ] is used to denote the empty list and a special function symbol, the dot '.' is used to denote an operator that may already be familiar as the *cons* function of Lisp, that is the function that takes a term **T** and a list **L** as arguments and returns the list whose first element is **T** and the rest of which, the so-called tail of the list, is **L**. Thus, by convention in Prolog, one thinks of the term **.(a,.(b,[ ]))** as denoting the list whose head is **a** and whose tail is the list consisting of the single element **b**. To make programs more readable the notation **[a,b]** is allowed for this but it should be emphasized that this is only for readability. There are no special 'list processing facilities' or 'list structures' in Prolog. The only data type is the term.

A basic operation in many computations is a test for set membership. Again we have to choose some way of representing sets by terms. One possibility is to represent sets as lists, and if we do this we can program set membership in Prolog as

```
member(X,[H|T]):-X=H.
member(X,[H|T]):-member(X,T).
```

This introduces some more notation. **[H|T]** is allowed as a more readable synonym for the term **.(H,T)**, that is the list whose head is **H** and whose tail is **T**. The declarative reading of the first clause is that 'for all **X**, **H** and **T**, **X** is a member of the list with head **H** and tail **T** if **X** and **H** are the same term'. The second clause reads 'for all **X**, **H** and **T**, **X** is a member of the list with head **H** and tail **T** if **X** is a member of the list **T**'. Both of these statements are mathematically correct statements about the member relation. We show now that they are also operationally effective when executed by the Prolog interpreter in the manner described above.

One way in which the **member** program could be used is by typing the goal

```
?- member(b,[a,b]).
```

which would get the answer **yes** as a result of the following sequence of operations. As before, the goal is matched against the sequence of program clauses starting at the top. The goal **member(b,[a,b])** does not match **member(X,[H|T])** because **X** is instantiated to **a** and **H** to **b**, and **a=b** fails, so a match with the second clause is attempted. This succeeds with **X** instantiated to **b**, **H** to **a** and **Y** to **[b]**, and the body of the clause is now executed in an environment containing these variable bindings. So **member(b,[b])** is matched against the sequence of program clauses starting again at the top. This time it matches with the first because **X** and **H** are both instantiated to **b** and so **X = H** succeeds and hence the original goal also succeeds.

Logically it is irrelevant, but as a point of programming style most Prolog programmers would prefer to write the two clauses above as

```
member(X,[X|_]).
member(X,[_|Y]):-member(X,Y).
```

where the '_' symbol is used to denote a variable whose binding is irrelevant. Note also how the **X=H** in the first clause is unnecessary because it can be incorporated by replacing **H** by **X** in the clause head. Another way to write the definition of **member** is

```
member(X,[Y|Z]):-X=Y; member(X,Z).
```

where the semicolon symbol ';' is read as a logical 'or'. The declarative reading is thus 'For all **X**, **X** is a member of the list whose head is **Y** and whose tail is the list **Z** if, either **X** is the same as **Y**, or **X** is a member of **Z**'. Logically and operationally all these definitions of member are identical. Neatness, readability and personal preference determine which style is used in practice.

Another basic relation on lists is **concatenation**, which often goes by the name of **append** in Prolog systems, and can be implemented by the following clauses

```
append([],Y,Y).
append([H|X],Y,[H|Z]):- append(X,Y,Z).
```

Once again these clauses correspond directly to the base case and induction step in proofs of statements about the **append** relation. As for **member**, there is more than one way to use **append**. For example

```
?- append([a,b],[c,d],[a,b,c,d]).
   yes

?- append([a,b],[b,c],Z).
   Z=[a,b,b,c]
```

As an exercise it is a good test of understanding to go, as we did for **member**, through the steps that the Prolog interpreter has to make in order to produce these responses. A full explanation of **append** is given in the recommended textbooks. It is worth noting here though that **append** can be used in a variety of other ways and many useful predicates can be defined neatly, if not always efficiently, in terms of it. For example

```
member(X,Y):- append(_,[X|_],Y).
```

which says that **X** is a member of **Y** if there is a list with head **X** which, when appended to some other list, gives **Y**. The second argument of **append** is not fully instantiated here, and this brings up another of the ways in which Prolog differs radically from functional languages. With functions there is a distinction between domain and range, whereas with relations no such distinction can be made between the arguments. This has the consequence that many Prolog programs that can be used to compute the value of a function can also be used, unchanged, to compute its inverse. For example

```
?- append([a,b],Y,[a,b,c,d]).
   Y=[c,d]

?- append(X,Y,[a,b,c,d]).
   X=[]
   Y=[a,b,c,d];
```

```
X=[a]
Y=[b,c,d];
```

This ability to use a deterministic program 'backwards' to generate all solutions that satisfy a constraint is an extremely useful feature, but it is not a universal property of Prolog programs and examining its correctness often involves non-logical, operational reasoning.

If we are implementing set union rather than concatenation of lists then we have to remove duplicates. The **union** predicate can be defined with the clauses

```
union([],Y,Y).
union([H|X],Y,Z):- member(H,Y), union(X,Y,Z).
union([H|X],Y,[H|Z]):- not(member(H,Y)),union(X,Y,Z).
```

which again correspond to mathematically correct statements about a relation **union(X,Y,Z)** which is satisfied when the list represented by the term **Z** is the union of the lists represented by the terms **X** and **Y**. Note that we have carefully said 'list' rather than 'set' here. The definition given is adequate for computing the union of two sets, with a goal such as

```
?- union([a,b],[b,c],Z).
Z=[a,b,c]
yes
```

but a goal such as

```
?- union([a,b,],[b,c], [c,b,a]).
```

would give the result **no**, which is right for lists but wrong for sets. There is nothing in the program to say that two lists that are permutations of each other represent the same set. One way to do this is given below. We leave the reader to work out how to use this in the **union** program.

```
same_set([H|U],V):- take_out(H,V,W),same_set(U,W).
same_set([],[]).
take_out(H,[H|T],T):-!.
take_out(X,[H|T],[H|U]):- take_out(X,T,U).
```

The usual close correspondence between recursion and mathematical induction is again evident in the **union** program. The induction is on the size of the list corresponding to the first argument. The base case is given by the first clause which says that the union of **Y** with the null list is **Y**. The second and third clauses deal with the two possible cases of the induction step.

Taking an element **H** of the first list: either **H** is a member of **Y**, in which case the relation is satisfied if **Z** is the union of **Y** and the remainder of the first list **X**, or **H** is not a member of **Y**, in which case the union is a list containing both **H** and **Z** where **Z** is the union of **X** and **Y**.

These are clearly mathematically correct statements. Readers should satisfy themselves that they are operationally effective by working out, in the same way as we did above for **member**, what happens when, for example, the goal **?-union([a,b],[b,c],Z)** is satisfied to give the answer substitution **Z=[a,b,c]**. Before you can do this we have to explain the system-defined predicate **not**. The goal **not(X)** is defined to succeed if the term that **X** is instantiated to fails when an attempt is made to satisfy it as a goal. So in this case to see, when some goal matches the third clause, whether **not(member(H,Y))** succeeds we have to see whether **member(H,Y)** succeeds, taking account of what **H** and **Y** might at that point be instantiated to. If **member(H,Y)** succeeds then **not(member(H,Y))** fails, but if **member(H,Y)** fails then **not(member(H,Y))** succeeds. This definition of **not** is known as **negation-as-failure** and it will be seen, certainly after reading Chapter 3, that it is not the same, in general, as logical negation. Being unable to prove P in some system of deduction is not meta-logically equivalent to being able to prove $\neg P$.

The second and third clauses of the **union** program are mutually exclusive in the sense that a non-empty set must match one or the other and cannot match both. In the case where an attempted match with the second clause fails because **H** is not a member of **Y** a completely redundant repetition of the **member** computation is made when the third clause is executed to show that **not(member(H,Y))** succeeds. One cannot simply leave out the **not(member(H,Y))** because the union program might be used in another program where backtracking forces the third clause to be tried after the second has succeeded, and this would be mathematically unsound. To get round this Prolog interpreters provide a system predicate denoted by the symbol '!'; and pronounced 'cut'. The example above would be rewritten using cut as

```
union([],Y,Y).
union([H|X],Y,Z):- member(H,Y),!,union(X,Y,Z).
union([H|X],Y,[H|Z]):- union(X,Y,Z).
```

The cut predicate always succeeds and has the side-effect of removing all choices available up to that point on subsequent backtracking. So, in the example above, if the second clause of **union** is being tried, and **member(H,Y)** succeeds, then the cut succeeds and removes the possibility of backtracking to the third clause.

One sometimes sees use of the cut vilified as a dangerously operational, non-logical notion, but properly used it is perfectly safe and can make programs more concise as well as more efficient. A common construction, which arises in many of the examples in this book, is the Prolog equivalent of the if–then–else construction of other languages. The general form is

```
p :- q,!,r;s.
```

which says that **p** is satisfied either if **q** is satisfied and **r** is satisfied, or if **q** is not satisfied but **s** is satisfied (that is, if **q** then **r** else **s**).

There are many other ways to use the cut to manipulate the operational behaviour of Prolog programs. Sterling and Shapiro give a discussion of the issues. From a pragmatic point of view if there is any doubt about the correctness of a construction involving 'cut' then it should not be used – there will always be a corresponding definition using '**not**'. However, one or other of the constructions is often unavoidable. There are good mathematical reasons why it is not possible to write clauses defining set intersection, for example, without using negation or some logically equivalent construction.

A.3.3 Taking terms to pieces

System-defined predicates are provided for getting inside and manipulating terms. Of these the most useful for our purposes is one whose predicate symbol is **=..**, pronounced 'univ' for historical reasons. Used without any user-defined program this has the following behaviour.

```
?- f(a,b) = ..X.
   X = [f,a,b]

?- f(a,b) = ..[f | Args].
   Args = [a,b]

?- Term = ..[f,a,b].
   Term = f(a,b)

?- [a] = ..Functor_and_Args.
   Functor_and_Args = [.,a,[]]
```

The last example in particular is a good test of understanding. Recall that lists are terms of arity 2, whose function symbol is '.' In general **T =.. L** is satisfied if **L** is or can be instantiated to a list and **T** can be unified with the term whose function symbol is the head of **L** and whose sequence of arguments is the tail of **L**. We shall see in the examples below, and in the main text, that, even though function symbols in Prolog terms have to be ground, we can use **=..** to write programs which operate on general terms. So this is a departure from strict first-order logic, one of the **meta-logical** features of Prolog.

Other useful system predicates in the same category are **functor(T, F, N)** which unifies **F** and **N** with the function symbol and number of arguments respectively of the term to which **T** is instantiated, and **arg(I,T,X)** which, if **I** is instantiated to an integer **i**, unifies **X** with the **i**th argument of the term to which **T** is (and must be) instantiated. Thus

```
?- functor([a],F,N).
   F=.
   N=2

?- arg(2,[a],X).
   X=[]
```

Finally, a pragmatic point, it is worth noting that most Prolog systems implement '=..' in terms of **functor** and **arg** so the latter should be used if efficiency is at a premium.

A.3.4 Arithmetic

Even in algebraic computation we sometimes have to do some arithmetic. An arithmetic expression such as 2 + 3 is a term whose function symbol is + and whose arguments are 2 and 3. By convention in mathematics the function symbol '+' is infixed, that is written between its two arguments, and the same facility is provided in Prolog for function and predicate symbols to be defined as infixed if required – we have already seen this in the case of the system predicates = and =... If, for example, we give the interpreter the goal

```
?- 2+3=..X.
```

then we get the output **X** = [+,2,3], showing that **2** + **3** is recognized by the system as a term whose function symbol is + and whose arguments are the constants **2** and **3**. Because **2+3** is just a term like **f(2,3)** the goal

```
?- X=2+3.
```

gives the, at first sight unexpected, answer **X=2+3**. This is because in first-order logic there is nothing to say that a term represents a function that may be applied to arguments to yield a value. It is perfectly feasible by purely logical means to define a predicate that can be used to evaluate an expression, but all Prolog systems in fact provide a system predicate '**is**' for this purpose. So

```
?- X is 2+3.
```

succeeds with **X** instantiated to the result of evaluating the right-hand side by the normal conventions of arithmetic, giving in this case the output **X=5**. It is important to understand the difference between = which succeeds if both sides can be unified and '**is**' which succeeds if the right-hand side can be evaluated arithmetically and then both sides unified. Again more detail is given in the recommended Prolog texts.

A.3.5 User-defined operators

We have seen that the arithmetic operators can be used in the conventional infixed form. We now go on to show that programmers can define their own function symbols as infixed. To give an example, the goal

```
:- op(550,xfx,[ –> ]).
```

succeeds and has the side-effect of recording that the compound '–>' is an infixed operator with precedence 550 and associativity of type '**xfx**'. Precedence and associativity are used to disambiguate terms whose structure is not made explicit through the use of parentheses. Again this is covered in full detail in the recommended books; here we just give a few examples.

The precise value of the precedence number is of no significance; it is only used as a convenient way to specify a linear ordering. The general rule is that the function symbol of highest precedence is the principal connective. So if we also include in the program

```
:- op(540,yfx,[/\]).
```

then we are saying that the string **a/\b –> c** stands for the term **(a/\b) –> c** rather than **a/\(b –> c)**. The second argument of the **op** predicate, the associativity type, can be used to say, for example, that **a – b – c** stands for **(a – b) – c** rather than **a – (b – c)**, by including at the head of the program the goal

```
:- op(500,yfx,[ – ]).
```

although this would, in fact, be unnecessary since most Prolog systems already include standard definitions for the arithmetic operators. Note that these operator definitions have to be included as goals rather than definitions (:- is essentially the same as ?-) because they have to be satisfied, and not just stored, in order to produce the required side-effect. For the same reason they must appear before the operators that they introduce are used. For present purposes we need these operator definitions in order to be able to use the conventional logical notation in our programs (see also Section 2.1.2). For the examples in this book we have used the definitions

```
:- op(510, fx,     [~]).
:- op(520,yfx,     [/\]).
:- op(530,yfx,     [\/]).
:- op(540,xfx,   [–>]).
:- op(550,xfx, [<–>]).
:- op(560,xfx,      [?]).
```

for the logical connectives and the meta-logical 'turnstile' ⊢, but other choices for the symbols would be perfectly valid. It cannot be emphasized too often that, to the Prolog interpreter, expressions involving these symbols are just terms, they have no logical significance.

A.3.6 Some common patterns in manipulating terms

We conclude this brief introduction to Prolog by discussing two of the basic program schemas that arise in manipulating terms.

A common pattern of recursion is illustrated by the following program. A term **S** is a subterm of a term **T** either if it is identical to **T** or if it is a subterm of one of the arguments of **T**. In the latter case we use the =.. system predicate to access the list of arguments of **T** and, if necessary, to check **S** against each of them.

```
subterm(T,T).
subterm(S,T) :- T = ..[_| Args], sub(S,Args).
sub(S,[First | Rest]) :- subterm(S,First); sub(S,Rest).
```

Note the use of the subsidiary predicate **sub** to process the list of arguments. It would be wrong to use **subterm(S,Args)** at the end of the second line because **subterm** expects a general term and not just a list as its second argument. Note also that here we do not need a base case for **sub** because we want it to fail if the second argument is the null list [ ]. Smaller points of style are the use of the anonymous variable '_' in the second line because the precise value of the function symbol does not matter here, and the use of the 'or' construction with ';' which seems neater here than having two clauses, one for the case where **S** is a subterm of the first argument, and the other for the recursive case.

Another frequently occurring type of program is where we have a different case for each operator in some algebra. We could illustrate this with the logical connectives defined above but there are several such examples in the text. Suppose instead that we want to define predicates to evaluate expressions in an algebra of sets with two binary infixed operators \/ and /\ denoting union and intersection and a unary prefixed operator ~ denoting complement with respect to some given universal set. The program, headed by the appropriate operator definitions, might be

```
:- op(500,  fx,[ ~ ]).
:- op(510,xfx,[/\]).
:- op(520,xfx,[\/]).

set([]).
set([_| S]) :- set(S).
universal_set([a,b,c,d]).

complement([],_,[]).
complement([H | T],X,Y) :- complement(T,X,Z),
                          member(H,X),!,Y = Z; Y = [H | Z]).
```

```
value(A\/B,C):- value(A,U),value(B,V),union(U,V,C).
value(A/\B,C):- value(A,U),value(B,V),intersection(U,V,C).
value( ~A,C):- value(A,B),universal_set(U),complement(U,B,C).
value(A,A):- set(A).
```

where **union** is given above and **intersection** (left as an exercise) is very similar to **union**.

The program would be used by typing goals of the form

```
?- X=[a,b],Y=[b,c],value(~((X/\~Y)\/(Y/\~X)),Z).
```

when the appropriate answer **Z=[b,d]** should be output.

Demands on space preclude any more examples, but there are plenty in the text. Some are more complicated, but all employ the basic ideas covered here. The aim of this brief introduction to algebraic manipulation in Prolog has been to prepare the ground for using and understanding them.

APPENDIX B

Programs in Standard ML and Prolog

B.1 Programs in SML

B.1.1 A parser for propositional logic

```
(* The datatype that represents sentences of propositional logic internally *)

datatype SENT = Null |
                Prop of string |
                Not of SENT |
                And of (SENT * SENT) |
                Or of (SENT * SENT) |
                Imp of (SENT * SENT) |
                Eq of (SENT * SENT);

        (* Some useful functions *)

fun eq a (b:string) = (a = b);
fun fst (a,b) = a;

        (* Variables are lowercase letters a to z * )

fun var h = "a" <= h andalso h <= "z";
fun andoper oper = (oper = "&") orelse (oper = "V");
fun impoper oper = (oper = " > ") orelse (oper = " = ");

exception varerror;
fun makeVar (h, rest) = if var h
                        then (Var h, rest)
                        else raise varerror;
```

```
exception error;
fun rest P (head::tail) = if P head
                          then tail
                          else raise error;

        (* The parser from strings to SENT *)

fun constructsent str = auxconstructsent(constructterm str)
and auxconstructsent (l, nil) = (l, nil)
  | auxconstructsent (l, head::tail) =
          if impoper head
          then let val (r, tail) = constructterm tail in
                   auxconstructsent(if head = ">"
                                    then (Imp(l,r), tail)
                                    else (Eq(l,r) ,tail))
               end
          else (l, head::tail)
and constructterm tail = auxconstructterm(constructfact tail)
and auxconstructterm (l, nil) = (l, nil)
  | auxconstructterm (l, head::tail) =
                   if andoper head
                   then let val (r,tail) =
                            constructfact tail in
                              auxconstructsent(if head = "&"
                                               then (And(l,r), tail)
                                               else (Or(l,r), tail))
                        end
                   else (l, head::tail)
and constructfact nil = raise error
  | constructfact (head::tail) =
                     if head = "~"
                     then let val (t1, tail) =
                                   constructfact tail in auxconstructsent (Not(t1),tail)
                          end
                     else if head = "("
                          then let val (t1,tail) = constructsent tail in
                                   if hd tail = ")"
                                   then (t1, rest (eq ")") tail)
                                   else raise error
                               end
                          else makeVar (head , tail);

      (* Packaging the parser to make it easily useable *)

fun strip nil = nil
  | strip S = if hd S = " "
              then strip (t1 S)
              else (hd S) :: (strip (t1 S));

fun make S = fst(constructsent (strip (explode S)));
```

```
        (* Unparsing, ready for printing *)

fun printsent (Var v) = v
  | printsent (And(left, right)) = "(" ^ printsent left ^ "&" ^ printsent right ^ ")"
  | printsent (Not(rest)) = "(" ^ "~" ^ printsent rest ^ ")"
  | printsent (Or(left, right)) = "(" ^ printsent left ^ "V" ^ printsent right ^ ")"
  | printsent (Imp(left, right)) = "(" ^ printsent left ^ ">" ^ printsent right ^ ")"
  | printsent (Eq(left, right)) = "(" ^ printsent left ^ "=" ^ printsent right ^ ")";

(* Finally, we parse some input and print it out if it is correct *)

val pr = printsent o make;
```

B.1.2 Extending the parser for predicate calculus

```
datatype TERM = Var of string |
                Name of string |
                app of (string * TERM list) |
                empty;

datatype SENT = Null |
                Prop of string |
                Pred of (string * TERM list) |
                Not of SENT |
                And of (SENT * SENT) |
                Or of (SENT * SENT) |
                Imp of (SENT * SENT) |
                Eq of (SENT * SENT) |
                Forall of (TERM * SENT) |
                Exists of (Term * SENT);

        (* Some useful functions *)

fun eq a (b:string) = (a = b);
fun fst (a,b) = a;

        (* Predicate letters are capitals from M to T *)

fun proporpred h = "M" <= h andalso h <= "T";
fun andoper oper = (oper = "&") orelse (oper = "V");
fun impoper oper = (oper = ">") orelse (oper = "=");

(* Names start with lower case letters from a to e ... *)

fun name n = "a" <= n andalso n <= "e";
        (* ...and carry on with one or more digits *)
fun number n = "0" <= n andalso n <= "9";
        (* Variables are lower case letters from x to z *)
fun variable v = "x" <= v andalso v <= "z";
        (* Functions letters are lower case from f to h *)
fun function f = "f" <= f andalso f <= "h";
fun quantifier q = (q = "A") orelse (q = "E");
```

```
exception Var_error
and       Name_error
and       term_error1
and       term_error2
and       app_error1
and       app_error2
and       app_error3
and       termlist_error0
and       termlist_error1
and       termlist_error2;

fun bldVar (v,nil) = raise Var_error
  | bldVar (v, h::t) = if number h
                          then bldVar(v ^ h,t)
                          else (Var v, h::t);

fun bldName (n,nil) = raise Name_error
  | bldName (n,h::t) = if number h
                          then bldName(n ^ h, t)
                          else (Name n,h::t);
fun bldterm nil = raise term_error1
  | bldterm (h::t) = if name h
                        then bldName (h,t)
                        else if variable h
                              then bldVar(h,t)
                              else if function h
                                    then bldapp(h,t)
                                    else raise term_error2
and bldapp (f,nil) = raise app_error1
  | bldapp (f,h::t) = if number h
                         then bldapp(f ^ h,t)
                         else let val (terms,t') = bldtermlist (h::t) in
                                   if terms = nil
                                   then raise app_error2
                                   else (app(f,terms),t')
                              end
and bldtermlist (h::t) = if h = "("
                            then let val (term,t') = bldterm t in
                                      bldtermlist' ([term],t')
                                 end
                            else raise termlist_error0
and bldtermlist' (a,nil) = raise termlist_error1
  | bldtermlist' (t1,h::t) = if h = ")"
                                then (t1,t)
                                else if h = ","
                                      then let val (term,t') = bldterm t in
                                              bldtermlist'(t1@[term],t')
                                           end
                                      else raise termlist_error2;
```

```
exception ProporPred_error;
fun makeProporPred (h, rest) = if proporpred h
                               then makePorP(h,rest)
                               else raise ProporPred_error
and makePorP (p,nil) = (Prop p,nil)
  | makePorP (p,h::t) = if number h
                        then makePorP(p ^ h,t)
                        else if h = "("
                             then let val (terms,rest) =
                                          bldtermlist (h::t) in
                                            (Pred(p,terms),rest)
                                  end
                             else (Prop p,h::t);

exception rest_error
and       fact_error0
and       fact_error1
and       quant_error0;

fun rest P (head::tail) = if P head
                          then tail
                          else raise rest_error;

fun constructsent str = auxconstructsent(constructterm str)
and auxconstructsent (l, nil) = (l, nil)
  | auxconstructsent (l, head::tail) =
                        if impoper head
                        then let val (r, tail) = constructterm tail
                             in auxconstructsent(
                                         if head = ">"
                                         then (Imp(l,r),tail)
                                         else (Eq(l,r),tail))
                              end
                        else (1, head::tail)
and constructterm tail = auxconstructterm(constructfact tail)
and auxconstructterm (l, nil) = (l, nil)
  | auxconstructterm (l, head::tail) =
                             if andoper head
                             then let val (r,tail) = constructfact tail in
                                                auxconstructterm(
                                                if head = "&"
                                                then (And(l,r),tail)
                                                else (Or(l,r),tail))
                                  end
                             else (l, head::tail)
and constructfact nil = raise fact_error0
  | constructfact (head::tail) =
                   if head = "~"
                   then let val (t1, tail) = constructfact tail in
                                      auxconstructsent (Not(t1),tail)
                        end
```

```
                    else if head = "("
                        then let val (t1, tail) = constructsent tail in
                                 if hd tail = ")"
                                 then (t1, rest (eq ")") tail)
                                 else raise fact_error1
                             end
                        else if quantifier head
                             then bldquant (head,tail)
                             else makeProporPred (head,tail)
and bldquant (h,t) =
                  let val (v,tail) = bldVar (hd t,tl t) in
                     let val (S,tail') = constructfact tail in
                        if h = "A"
                        then auxconstructsent(Forall(v,S),tail')
                        else if h = "E"
                             then auxconstructsent(Exists(v,S),tail')
                             else raise quant_error0
                     end
                  end;

fun strip nil = nil
  | strip S = if hd S = " "
              then strip (tl S)
              else (hd S)::(strip (tl S));

(* This is the final function that maps a string to a SENT *)

fun make S = fst(constructsent (strip (explode S)));

(* Auxiliary functions for printing terms, used in... *)

exception printterms_error0;
fun printterms nil = raise printterms_error0
  | printterms ((Name n)::t) = "(" ^ n ^ (printterms' t)
  | printterms ((Var v)::t) = "(" ^ v ^ (printterms' t)
  | printterms ((app(f,terms))::t) = "(" ^ f ^ (printterms terms) ^ (printterms' t)
and printterms' nil = ")"
  | printterms' ((Name n)::t) = "," ^ n ^ (printterms' t)
  | printterms' ((Var v)::t) = "," ^ v ^ (printterms' t)
  | printterms' ((app(f,terms))::t) = "," ^ f ^ (printterms terms) ^ (printterms' t);

(* ...this function, which takes SENTs and prints them as strings *)

fun printsent (Prop v) = v
  | printsent (And(left, right)) = "(" ^ printsent left ^ "&" ^ ")" printsent right ^ ")"
  | printsent (Not(rest)) = "(" ^ "~" ^ printsent rest ^ ")"
  | printsent (Or(left, right)) = "(" ^ printsent left ^ "V" ^ printsent right ^ ")"
  | printsent (Imp(left, right)) = "(" ^ printsent left ^ ">" ^ printsent right ^ ")"
  | printsent (Eq(left, right)) = "(" ^ printsent left ^ "=" ^ ")" printsent right ^ ")"
  | printsent (Pred(p,terms)) = p ^ printterms terms
  | printsent (Forall(var v,S)) = "A" ^ v ^ "(" ^ printsent S ^ ")"
  | printsent (Exists(Var v,S)) = "E" ^ v ^ "(" ^ printsent S ^ ")";
```

B.1.3 Using truth-tables efficiently

```
(* Here is a collection of functions to use the truth-tables method for deciding validity of
entailments. Instead of building the whole truth-table we check line by line because we
can stop whenever the conditions for validity are first broken, rather than going through all
possibilities *)

fun filter p nil = nil
  | filter p (x::xs) = if p x
                       then x::filter p xs
                       else filter p xs

fun mem nil a = false
  | mem (x::xs) a = a = x orelse mem xs a;

fun union(s,t) = s @ filter (not o mem s) t;

fun apply((s,v)::rest, S) = if s = S
                            then v
                            else apply(rest,S);

exception finished;
fun switch' nil a = raise finished
  | switch' ((sent,true)::rest) a = a@((sent,false)::rest)
  | switch' ((sent,false)::rest) a = switch' rest ((sent,true)::a);
fun switch a = switch' a nil;

fun set_up (Var s) = [(s,true)]
  | set_up(Not s) = set_up(s)
  | set_up(And(s,t)) = union(set_up(s),set_up(t))
  | set_up(Or(s,t)) = union(set_up(s),set_up(t))
  | set_up(Imp(s,t)) = union(set_up(s),set_up(t))
  | set_up(Eq(s,t)) = union(set_up(s),set_up(t));

local fun valid' (s,valuation) =
                      if (truthvalue valuation s)
                      then output std_out "\nincorrect\n"
                      else valid'(s,switch valuation) in
      fun valid sent = valid'(sent,set_up sent)
end;

fun transform(nil,goal) = Not(goal)
  | transform(x::nil,Null) = x
  | transform(x::xs,goal) = And(x,transform(xs,goal));

infix ttentails;

        (* Here is the main function.
               Given a list of assumptions ass and
                      a conclusion conc we evaluate
                                   ass ttentails conc *)

fun ass ttentails conc =
        (valid(transform(ass,conc)))
        handle finished => output std_out "\ncorrect\n");
```

B.1.4 Proofs using tableaux

```
(* Here is a version of the tableau algorithm which shortens
                    the amount of tableau traversal by keeping a
                        note of all the sentences passed on the way
                            from root to node. This makes testing
                                                  for closure much quicker *)

type nodeinfo = {Closed : bool, Used : bool};
datatype Tableau = Leaf of (SENT * nodeinfo) |
                     Onenode of (SENT * nodeinfo * Tableau) |
                     Twonode of (SENT * nodeinfo * Tableau * Tableau);

fun Neg (Not(sent)) = sent | Neg sent = Not(sent);
infix mem;
fun a mem nil = false | a mem (x::xs) = a = x orelse a mem xs;

val initinfo = {Closed = false,Used = false};
fun initialise (nil,goal) = Leaf(Neg(goal),initinfo)
  | initialise (premise::nil,Null) = Leaf(premise,initinfo)
  | initialise (premise::rest,goal) = Onenode(premise,initinfo,initialise (rest,goal));

fun alpha(And(s,t)) = true
  | alpha(Not(Imp(s,t))) = true
  | alpha(Not(Or(s,t))) = true
  | alpha((s) = false;

fun composite(Var s) = false
  | composite(Not(Var s)) = false
  | composite(s) = true;

fun apply_rule (Imp(s,t)) = (Neg(s),t)
  | apply_rule (Eq(s,t)) = (And(s,t),And(Neg(s),Neg(t)))
  | apply_rule(And(s,t)) = (s,t)
  | apply_rule(Or(s,t)) = (s,t)
  | apply_rule(Not(Imp(s,t))) = (s,Neg(t))
  | apply_rule(Not(Eq(s,t))) = (And(Neg(s),t),And(s,Neg(t)))
  | apply_rule(Not(And(s,t))) = (Neg(s),Neg(t))
  | apply_rule(Not(Or(s,t))) = (Neg(s),Neg(t));

val useup = {Closed = false,Used = true};
val useable = (fn a => fn b:nodeinfo => not( #Used(b)) andalso composite(a));
fun extend(Leaf(S,N),sent,path) =
      let fun test s = {Closed = Neg(s) mem path,Used = false}in
                 if not( #Closed(N))
                 then let val (f,s) = apply_rule(sent) in
                        if alpha(sent)
                        then Onenode(S,N,Onenode(f,test f,Leaf(s,test s)))
                        else Twonode(S,N,next'(Leaf(f,test f),S::path),
                                        next'(Leaf(s,test s),S::f::path))
                      end
                 else (Leaf(S,N))
      end
```

```
  | extend(Onenode(S,N,T),sent,path) =
                                  if not( #Closed(N))
                                  then Onenode(S,N,extend(T,sent,S::path))
                                  else Onenode(S,N,T)
  | extend(Twonode(S,N,T1,T2),sent,path) =
                                  if not( #Closed(N))
                                  then Twonode(S,N,extend(T1,sent,S::path),
                                                   extend(T2,sent,S::path))
                                  else Twonode(S,N,T1,T2)

and next' (Leaf(S,N),path) =
                                  if (useable S N)
                                  then extend(Leaf(S,useup),S,path)
                                  else Leaf(S,N)
  | next' (Onenode(S,N,T),path) =
                                  if (useable S N)
                                  then extend(Onenode(S,useup,T),S,path)
                                  else Onenode(S,N,next'(T,S::path))
  | next' (Twonode(S,N,T1,T2),path) =
                                  if (useable S N)
                                  then extend(Twonode(S,useup,T1,T2),S,path)
                                  else Twonode(S,N,next'(T1,S::path),next'(T2,S::path))
and next T = next'(T,nil);

fun closed (Leaf(S,N)) = #Closed(N)
  | closed (Onenode(S,N,T)) = #Closed(N) orelse closed T
  | closed (Twonode(S,N,T1,T2)) = #Closed(N) orelse ((closed T1)
                                                  andalso (closed T2));

fun make_tableau T = let val T' = next T in
                                  if T' = T
                                  then T
                                  else make_tableau T'
                          end;

infix entails;
fun asslist entails goal =
        let val start_tableau = initialise(asslist,goal) in
           let val final_tableau = make_tableau(start_tableau) in
              if closed(final_tableau)
              then output std_out "\ncorrect\n"
              else output std_out "\nincorrect\n"
           end
        end;
```

B.1.5 Terms, substitutions

```
(* Here we introduce some terms and
      show how we can do substitutions on them.
          We also show how composition of substitutions
                                      can be implemented *)
```

```
datatype term = app of string * term list |
                var of string |
                const of string |
                empty;
fun lookup(x,nil) = empty
  | lookup(x,(y,t)::rest) = if x = y
                            then t
                            else lookup(x,rest);
fun applySubst (Subst,(app(f,Args))) = app(f,(Sub(Subst,Args)))
  | applySubst (Subst, (var x)) =
              let val NewTerm = lookup(var x,Subst) in
                if NewTerm = empty
                then var x
                else NewTerm
              end
  | applySubst (_, (const a)) = const a
  | applySubst (_,empty) = empty
and Sub (_,nil) = nil
  | Sub (Subst,FirstArg::RestofArgs) =
            applySubst (Subst,FirstArg)::Sub (Subst,RestofArgs);
fun combine(S,nil) = S
  | combine(Sub,(Var,_)::Rest) = let val S = combine(Sub,Rest) in
                                   if not(lookup(Var,Sub) = empty)
                                   then S
                                   else (Var,Var)::S
                                 end;
fun compose(nil,_) = nil
  | compose((Var,Term)::Rest,Sub2) =
                 let val NewTerm = applySubst(Sub2,Term)
                     and S = compose(Rest,Sub2) in
                       if Var = NewTerm
                       then S
                       else (Var,NewTerm)::S
                 end;
fun composition(Sub1,Sub2) = compose(combine(Sub1,Sub2),Sub2);
```

B.1.6 Program for semantic tableau method for predicate calculus

```
fun lookup(x,nil) = empty
  | lookup(x,(y,t)::rest) = if x = y then t else lookup(x,rest);
fun applySubst Subst (app(f,Args)) = app(f,map (applySubst Subst) Args)
  | applySubst Subst (Var x) = let val NewTerm = lookup(Var x,Subst) in
                                 if NewTerm = empty
                                 then Var x
                                 else NewTerm
                               end
```

```
  | applySubst _ (Name a) = Name a
  | applySubst _ empty = empty;
type nodeinfo = {Closed : bool, Used : bool};
datatype Tableau =
                Leaf of (SENT * nodeinfo) |
                Onenode of (SENT * nodeinfo * Tableau) |
                Twonode of (SENT * nodeinfo * Tableau * Tableau);

fun fst (a,b) = a;
fun snd (a,b) = b;

infix mem;
fun a mem nil = false | a mem (x::xs) = a = x orelse a mem xs;
infix -;
fun l1 - nil = l1
  | l1 - (h::t) = take(h,l1) - t
and take(h,nil) = nil
  | take(h,a::b) = if h = a then b else a::take(h,b);
fun anddist nil = true
  | anddist (h::t) = h andalso (anddist t);

fun flatten nil = nil
  | flatten (h::t) = h @ (flatten t);

fun filter nil = nil
  | filter (h::t) = if (h mem t) then filter t else h::(filter t);

val allnames = [Name "a",Name "b",Name "c",Name "d",Name "e",
                    Name "a1",Name "a2",Name "a3"];

fun name (Name n) = true
  | name (app(f,l)) = anddist (map name l)
  | name s = false;

fun alpha(And(s,t)) = true
  | alpha(Not(Imp(s,t))) = true
  | alpha(Not(Or(s,t))) = true
  | alpha(s) = false;

fun beta((Or(s,t)) = true
  | beta(Imp(s,t)) = true
  | beta(Not(And(s,t))) = true
  | beta(Eq(s,t)) = true
  | beta(Not(Eq(s,t))) = true
  | beta(s) = false;

fun gamma(Forall(v,s)) = true
  | gamma(Not(Exists(v,s))) = true
  | gamma(s) = false;

fun delta(Exists(v,s)) = true
  | delta(Not(Forall(v,s))) = true
  | delta(s) = false;
```

```
fun composite(Pred(p,t)) = false
  | composite(Not(Pred(p,t))) = false
  | composite (Prop p) = false
  | composite (Not(Prop p)) = false
  | composite (s) = true;

fun Neg (Not(sent)) = sent | Neg sent = Not(sent);

fun instance term (Forall(x,s)) = instance'(s,x,term)
  | instance term (Not(Exists(x,s))) = Neg(instance'(s,x,term))
  | instance term (Exists(x,s)) = instance'(s,x,term)
  | instance term (Not(Forall(x,s))) = Neg(instance'(s,x,term))

and instance'(And(s,t),x,newt) = And(instance'(s,x,newt),instance'(t,x,newt))
  | instance'(Or(s,t),x,newt) = Or(instance'(s,x,newt),instance'(t,x,newt))
  | instance'(Imp(s,t),x,newt) = Imp(instance'(s,x,newt),instance'(t,x,newt))
  | instance'(Eq(s,t),x,newt) = Eq(instance'(s,x,newt),instance'(t,x,newt))
  | instance'(Not(s),x,newt) = Not(instance'(s,x,newt))
  | instance'(Forall(Var y,s),x,newt) =
                    if x = Var y
                    then Forall(Var y,s)
                    else Forall(Var y,instance'(s,x,newt))
  | instance'(Exists(Var y,s),x,newt) =
                    if x = Var y
                    then Exists(Var y,s)
                    else Exists(Var y, instance'(s,x,newt))
  | instance'(Pred(p,terms),x,newt) =
                    Pred(p,map (applySubst [(x,newt)]) terms)
  | instance'(p,x,newt) = p;

fun apply_rule (Imp(s,t)) = (Neg(s),t)
  | apply_rule (Eq(s,t)) = (And(s,t),And(Neg(s),Neg(t)))
  | apply_rule (And(s,t)) = (s,t)
  | apply_rule (Or(s,t)) = (s,t)
  | apply_rule (Not(Imp(s,t))) = (s,Neg(t))
  | apply_rule (Not(Eq(s,t))) = (And(Neg(s),t),And(s,Neg(t)))
  | apply_rule (Not(And(s,t))) = (Neg(s),Neg(t))
  | apply_rule (Not(Or(s,t))) = (Neg(s),Neg(t))
  | apply_rule p = (Null,Null);

fun newname nl = if (allnames - nl) = nil
                    then (Name "*")
                    else hd(allnames - nl);

fun getnames (And(s,t)) = (getnames s) @ (getnames t)
  | getnames (Or(s,t)) = (getnames s) @ (getnames t)
  | getnames (Imp(s,t)) = (getnames s) @ (getnames t)
  | getnames (Eq(s,t)) = (getnames s) @ (getnames t)
  | getnames (Not(s)) = getnames s
  | getnames (Forall(v,s)) = getnames s
  | getnames (Exists(v,s)) = getnames s
```

```
  | getnames (Pred(p,l)) = flatten (map getnames' l)
  | getnames p = nil

and getnames' (Name n) = [Name n]
  | getnames' (Var x) = nil
  | getnames' (app(f,l)) = (if name(app(f,l))
                              then [app(f,l)]
                              else nil) @ (flatten (map getnames' l))
  | getnames' empty = nil;

val initinfo = {Closed = false,Used = False};
fun initialise (nil,goal) = Leaf(Neg(goal),initinfo)
  | initialise (premise::nil,Null) = Leaf(premise,initinfo)
  | initialise (premise::rest,goal) = Onenode(premise,initinfo,initialise (rest,goal));

infix oneof;
fun S oneof (Leaf(S',N)) = S = S'
  | S oneof (Onenode(S',N,T)) = S = S' orelse (S oneof T);

fun initialclosed (Leaf(S,N)) = false
  | initialclosed (Onenode(S,N,T)) = (Neg(S) oneof T) orelse (initialclosed T);

fun nameson path = filter(flatten(map (fn (a,b) => getnames a) path));

infix onpath;
fun S onpath path = ((S,{Closed = true,Used = true})mem path) orelse
                    ((S,{Closed = true,Used = false}) mem path) orelse
                    ((S,{Closed = false,Used = true}) mem path) orelse
                    ((S,{Closed = false,Used = false}) mem path);

infix appears_unused_on;
fun S appears_unused_on path =
                 ((S,{Closed = true,Used = false}) mem path) orelse
                 ((S,{Closed = false,Used = false}) mem path);
fun test S s n path = {Closed = Neg(s) onpath (n::(S::path)), Used = false};

fun addgammanodes(S,N,sent,path) = agn(S,N,sent,path,nameson path)

and agn(S,N,sent,path,n::nil) =
                 let val newsent = instance n sent in
                       if newsent onpath path
                       then if sent appears_unused_on path
                             then Leaf(S,N)
                             else Onenode(S,N, Leaf(sent,initinfo))
                       else (print("\nuniversal\n" ^ (printsent newsent)^ "\n");
                             if sent appears_unused_on path
                             then Onenode(S,N,Leaf(newsent,test (S,N) newsent
                                          (Null,initinfo) path))
                             else Onenode(S,N,Onenode(newsent,test (S,N) newsent
                                          (Null,initinfo) path,Leaf(sent,initinfo))))
                 end
```

```
    | agn (S,N,sent,path,n::t) =
                    let val newsent = instance n sent in
                        if newsent onpath path
                        then agn(S,N,sent,path,t)
                        else (print ("\nuniversal\n" ^ printsent newsent ^ "\n");
                              if sent appears_unused_on path
                              then Onenode(S,N,agn(newsent,test (S,N)
                                          newsent (Null,initinfo) path,sent,path,t))
                              else Onenode(S,N,Onenode(newsent,test (S,N) newsent
                                          (Null,initinfo) path,agn(sent,test (S,N)
                                          newsent (Null,initinfo) path,sent,path,t))))
                    end
    | agn(S,N,sent,path,nil) =
                let val nn = instance (Name "a") sent in
                      (print ("\nuniversal\n" ^ (printsent nn)^"\n");
                       if sent appears_unused_on path
                       then Onenode(S,N,Leaf(nn, test (S,N) nn (Null,initinfo) path))
                       else Onenode(S,N,Onenode(nn, test (S,N) nn (Null,initinfo)
                                 path,Leaf(sent,initinfo))))
                end;

fun adddeltanode(S,N,sent,path) = adn(S,N,sent,nameson path,path)

and adn(S,N,sent,nl,path) =
        let val nn = instance (newname nl) sent in
           (print ("\nexistential\n" ^ (printsent nn)^ "\n");
            Onenode(S,N,Leaf(nn,test (S,N) nn (Null,initinfo) path)))
        end;

val useable = (fn a => fn b : nodeinfo => not( #Used(b)) andalso composite(a));

datatype kinds = prop | univ | exis | ground;

fun kindof s = if alpha(s) orelse beta(s)
               then prop
               else if gamma(s)
                    then univ
                    else if delta(s)
                         then exis
                         else ground;
val useup = {Closed = false,Used = true};

fun extend(Leaf(S,N),sent,path) kind =
                if not( #Closed(N))
                then case kind of
                     prop =>
                       if alpha(sent)
                       then let val (f,s) = apply_rule(sent) in
                                (Onenode(S,N Onenode(f,test (S,N) f
                       (Null,initinfo) path, Leaf(s,test (S,N) s (f,initinfo) path))),true)
                            end
```

```
                else if beta(sent)
                    then let val (f,s) = apply_rule(sent) in
                            (Twonode(S,N,Leaf(f,test (S,N) f (Null,initinfo) path),
                            Leaf(s,test (S,N) s (Null,initinfo) path)),true)
                        end
                    else (Leaf(S,N),false)
                  | univ =>
                    if gamma(sent)
                    then (addgammanodes(S,N,sent,(S,N)::path),true)
                    else (Leaf(S,N),false)
                  | exis =>
                    if delta(sent)
                    then (adddeltanode(S,N,sent,(S,N)::path),true)
                    else (Leaf(S,N),false)
                  | ground => (Leaf(S,N),false)
               else (Leaf(S,N),false)
  | extend(Onenode(S,N,T),sent,path) kind =
                        if not ( # Closed(N))
                        then let val (f,s) = extend(T,sent,(S,N)::path) kind in
                                                                (Onenode(S,N,F),s)
                             end
                        else (Onenode(S,N,T),false)
  | extend(Twonode(S,N,T1,T2),sent,path) kind =
                       if not( # Closed(N))
                       then let val (f1,s1) = extend(T1,sent,(S,N)::path) kind
                                and (f2,s2) = extend(T2,sent,(S,N)::path) kind in
                                                             (Twonode(S,N,f1,f2),s1 orelse s2)
                            end
                       else (Twonode(S,N,T1,T2),false)

and next' (Leaf(S,N),path) kind =
                       if (useable S N) andalso (kindof S) = kind
                       then extend(Leaf(S,useup),S,path) kind
                       else (Leaf(S,N),false)
  | next' (Onenode(S,N,T),path) kind =
                       if (useable S N) andalso (kindof S) = kind
                       then extend(Onenode(S,useup,T),S,path) kind
                       else let val (f,s) = next'(T,(S,N)::path) kind in
                                                                (Onenode(S,N,f),s)
                       end
  | next' (Twonode(S,N,T1,T2),path) kind =
                      if (useable S N) andalso (kindof S) = kind
                      then extend(Twonode(S,useup,T1,T2),S,path) kind
                      else let val (f1,s1) = next'(T1,(S,N)::path) kind
                               and (f2,s2) = next'(T2,(S,N)::path) kind in
                                             (Twonode(S,N,f1,f2),s1 orelse s2)
                         end

and propnext T = next'(T,nil) prop
and univnext T = next'(T,nil) univ
```

```
and exisnext T = next'(T,nil) exis;
fun closed (Leaf(S,N)) = #Closed(N)
  | closed (Onenode(S,N,T)) = #Closed(N) orelse closed T
  | closed (Twonode(S,N,T1,T2)) = #Closed(N) orelse ((closed T1)
                                                      andalso (closed T2));

fun make_tableau T = let val (T',changes) = propnext T in
                         if closed T'
                         then T'
                         else if changes
                              then make_tableau T'
                              else let val (T',changes) = exisnext T' in
                                       if closed T'
                                       then T'
                                       else if changes
                                            then make_tableau T'
                                            else let val (T',changes) = univnext T' in
                                                     if closed T'
                                                     then T'
                                                     else if changes
                                                          then make_tableau T'
                                                          else T'
                                                 end
                                   end
                     end;

infix entails;
fun asslist entails goal =
    let val start_tableau = initialise(map make asslist,make goal) in
        if initialclosed start_tableau
        then output std_out "\ncorrect\n"
        else let val final_tableau = make_tableau(start_tableau)in
                 if closed(final_tableau)
                 then output std_out "\ncorrect\n"
                 else output std_out "\nincorrect\n"
             end
    end;
```

B.2 Programs in Prolog

B.2.1 Using truth-tables

```
:- op(510, fx,  [~]).
:- op(520,xfy,  [/\]).
:- op(530,xfy,  [\/]).
:- op(540,xfx, [->]).
:- op(550,xfx,   [?]).
```

```
Assumptions?Goal :- transform(Assumptions,Goal,Formula),
                    setup(Formula,Valuation),
                    (generate(Valuation),
                    value(Formula,t,Valuation),
                    write('not valid'));
                    write('valid').

transform([],G,~G).
transform([H|T],G,H/\X) :- transform(T,G,X).

setup(A,[[A|_]]) :- atomic(A).
setup(~F,V) :- setup(F,V).
setup(F,V) :- F =..[_,A,B],setup(A,X),setup(B,Y),union(X,Y,V).

generate([]).
generate([[A,V]|T]) :- (V = t;V = f),generate(T).

value(     A,Z,V) :- atomic(A),!,member([A,Z],V).
value(   ~ A,Z,V) :- value(A,X,V),truth_table(~X,Z).
value(A /\ B,Z,V) :- value(A,X,V),value(B,Y,V),truth_table(X/\Y,Z).
value(A \/ B,Z,V) :- value(A,X,V),value(B,Y,V),truth_table(X\/Y,Z).
value(A -> B,Z,V) :- value(A,X,V),value(B,Y,V),truth_table(X -> Y,Z).

truth_table(t/\t,t) :- !.
truth_table(_/\_,f).
truth_table(f\/f,f) :- !.
truth_table(_\/_,t).
truth_table(t -> f,f) :- !.
truth_table(_ -> _,t).
truth_table(~t,f).
truth_table(~f,t).

union([],Y,Y).
union([[A,_]|T],Y,Z) :- member([A,_],Y),!,union(T,Y,Z).
union([[A,V]|T],Y,[[A,V]|Z]) :- union(T,Y,Z).
member(X,[H|T]) :- X = H;member(X,T).
```

Solutions to Selected Exercises

2.2

(b) The sentences in the arguments in (a) have the following truth-values:

sentence	*truth-value*
A	f
B	t
C	t
D	t
E	f

	premise	*conclusion*	*argument*
(a)	f	f	valid
(b)	t	t	invalid
(c)	t	f	invalid
(d)	t	t	valid
(e)	f	t	valid
(f)	f	f	invalid
(g)	f	t	invalid

This shows that an invalid argument can have

true premises and true conclusion
true premises and false conclusion
false premises and true conclusion
false premises and false conclusion

and that a valid argument can have

true premises and true conclusion
false premises and true conclusion
false premises and false conclusion.

(c) No such argument can be constructed since a valid argument cannot have true premises and false conclusion. This is because an argument is valid iff the sentence which is 'conjunction of its premises $\rightarrow$ its conclusion' is a tautology. That this was so would contradict the fact that the premises were true and the conclusion false.

2.3

(e) We use the abbreviations:

A: The Eiffel Tower is in Australia
B: Australia is below the equator
C: The Eiffel Tower is in Paris
D: Paris is in France
E: France is in Australia

(i)

$A \wedge B$
$\therefore A$

A	B	$A \wedge B$	$(A \wedge B) \rightarrow B$
t	t	t	t
t	f	f	t
f	t	f	t
f	f	f	t

The argument is valid since it is true under all valuations, that is lines of the truth-table.

(ii)

$C \vee D$
$\therefore C$

C	D	$C \vee D$	$(C \wedge D) \rightarrow C$
t	t	t	t
t	f	t	t
f	t	t	f
f	f	f	t

The argument is invalid since the third line makes the argument false, so it is not true under all valuations.

(iii)

$A \vee B$
$\therefore B$

This is the same form of argument as (b), so it is invalid too.

(iv)

$C \wedge D$
$\therefore C$

This is the same form of argument as (a), so it is valid too.

(v)

$A \wedge D$
$\therefore D$

As $A \wedge D$ is equivalent to $D \wedge A$, this is the same form of argument as (a), so it is valid too.

(vi)

$A \vee E$
$\therefore A$

This is the same form of argument as (c), so it is valid too.

(vii)

$A \vee E$
$\therefore C$

A	E	$A \vee E$	C	$(A \vee E) \rightarrow C$
t	t	t	t	t
t	t	t	f	f
t	f	t	t	t
t	f	t	f	f
f	t	t	t	t
f	t	t	f	f
f	f	f	t	t
f	f	f	f	t

Here, there are valuations, that is lines of the truth-table, where the argument is false, so it is invalid.

2.6

(d)

```
fun truthvalue V Null = raise invalid_sentence
  | truthvalue V (Var P) = apply(V,P)
  | truthvalue V (Not(S)) =
          if truthvalue V S = true
          then false else true
  | truthvalue V (And(S, T)) =
          if truthvalue V S andalso truthvalue V T = true
          then true else false
  | truthvalue V (Or(S, T)) =
          if truthvalue V S = true orelse truthvalue V T = true
          then true else false
  | truthvalue V (Imp(S, T)) =
          if truthvalue V S = true andalso truthvalue V T = false
          then false else true
  | truthvalue V (Eq(S, T)) =
          if truthvalue V S = truthvalue V T then true else false;
```

2.7

(c)

```
fun rhsimp(s,nil) = (false,Null)
  | rhsimp(s,(Imp(l,r))::t) = if s = r then (true,l) else rhsimp(s,t)
  | rhsimp(s,h::t) = rhsimp(s,t);
fun MP l s = let val (ok,left) = rhsimp(s,l) in ok andalso (left memberof l)
             end;
fun proof l = proof' nil l
and proof' l nil = true
  | proof' l (h::t) = (axiom h orelse MP l h) andalso (proof'(l @ [h]) t);
```

3.2

(d)

```
infix free_for;
fun t free_for (x, And(S,T)) = (t free_for (x,S)) andalso (t free_for (x,T))
    | t free_for (x ,Or(S,T)) = (t free_for (x,S)) andalso (t free_for (x,T))
    | t free_for (x ,Imp(S,T)) = (t free_for (x,S)) andalso (t free_for (x,T))
    | t free_for (x ,Eq(S,T)) = (t free_for (x,S)) andalso (t free_for (x,T))
    | t free_for (x ,Not(S)) = (t free_for (x,S))
    | t free_for (x ,Forall(v,S)) = not(x free_in (Forall(v, S)) andalso
                                  v occurs_in [t] andalso (t free_for (x,S))
    | t free_for (x ,Exists(v,S)) = not(x free_in (Exists(v,S)) andalso
                                  v occurs_in [t] andalso (t free_for (x,S))
    | _free_for_ = true;
```

(g) We first give new two auxiliary functions before we can give the definition of forms_match:

```
fun combine (b,t) (b't') = let val b'' = (b andalso b') in
                               if t = t'
                               then (b'',t)
                               else if t = empty
                                    then (b'',t')
                                    else if t' = empty
                                         then (b'',t)
                                         else (false,empty)
                           end;
fun terms_match (x,nil) nil = (true,empty)
  | terms_match (x,(Name n)::tail) ((Name n')::tail') =
            if n = n'
            then (terms_match (x,tail) tail')
            else (false,empty)
  | terms_match (x, (Var v)::tail) (t::tail') =
            if x = Var v
            then (combine (terms_match (x,tail) tail') (true,t))
            else if t = Var v
                 then (terms_match (x,tail) tail')
                 else (false,empty)
```

```
  | terms_match (x,(app(f,terms))::tail) ((app(f',terms'))::tail') =
              if f = f'
              then (terms_match (x,tail) tail')
              else (false, empty)
  | terms_match _ _ = (false,empty);
fun forms_match (x,And(S,T)) (And(S',T')) =
              combine (forms_match(x,S) S') (forms_match(x,T) T')
  | forms_match (x,Or(S,T)) (Or(S',T')) =
              combine (forms_match(x,S) S') (forms_match(x,T) T')
  | forms_match (x,Imp(S,T)) (Imp(S',T')) =
              combine (forms_match(x,S) S') (forms_match(x,T) T')
  | forms_match (x,Eq(S,T)) (Eq(S',T')) =
              combine (forms_match(x,S) S') (forms_match(x,T) T')
  | forms_match (x,Not(S)) (Not(S')) = (forms_match (x,S) S')
  | forms_match (x, Forall(v,S)) (Forall(v',S')) =
                    if v = v'
                    then (forms_match (x,S) S')
                    else (false,empty)
  | forms_match (x,Exists(v,S)) (Exists(v',S')) =
                    if v = v'
                    then (forms_match (x,S) S')
                    else (false,empty)
  | forms_match (x,Pred(p,terms)) (Pred(p',terms')) =
                    if p = p'
                    then (terms_match (x,terms) terms')
                    else (false,empty)
  | forms_match (x,Prop p) (Prop p') = (p = p',empty)
  | forms_match _ _ = (false,empty);
```

(h)

```
fun proof l = proof' nil l
and proof' l nil = true
  | proof' l (h::t) = (axiom h orelse MP l h
                      orelse generalization l h) andalso
                      (proof' (l @ [h]) t);
```

4.1

(a)

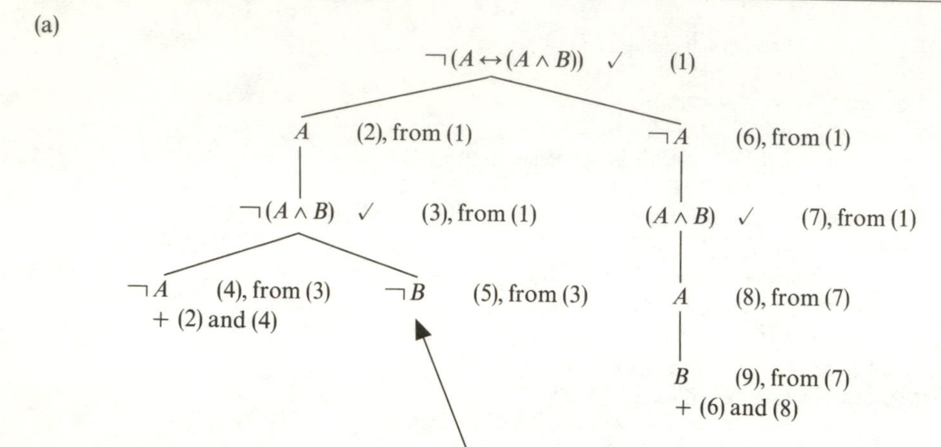

A	B	$A \wedge B$	$A \leftrightarrow (A \wedge B)$
t	t	t	t
t	f	f	f
f	t	f	f
f	f	f	f

← this is the only line where the conclusion is true so since it is not true in all valuations, that is lines of the truth-table, the **entailment is not valid**

(b)

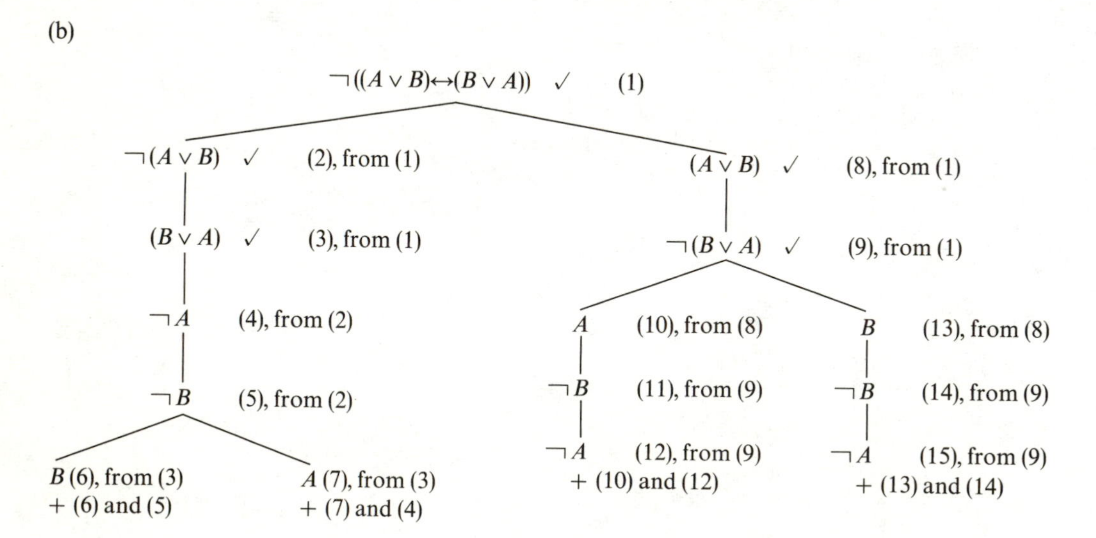

All paths are closed, so **entailment is valid**.

A	B	$A \vee B$	$B \vee A$	$(A \vee B) \rightarrow (B \vee A)$	
t	t	t	t	t	← the conclusion is true
t	f	t	t	t	← in all valuations
f	t	t	t	t	← and so the
f	f	f	f	t	← **entailment is valid**

(c)

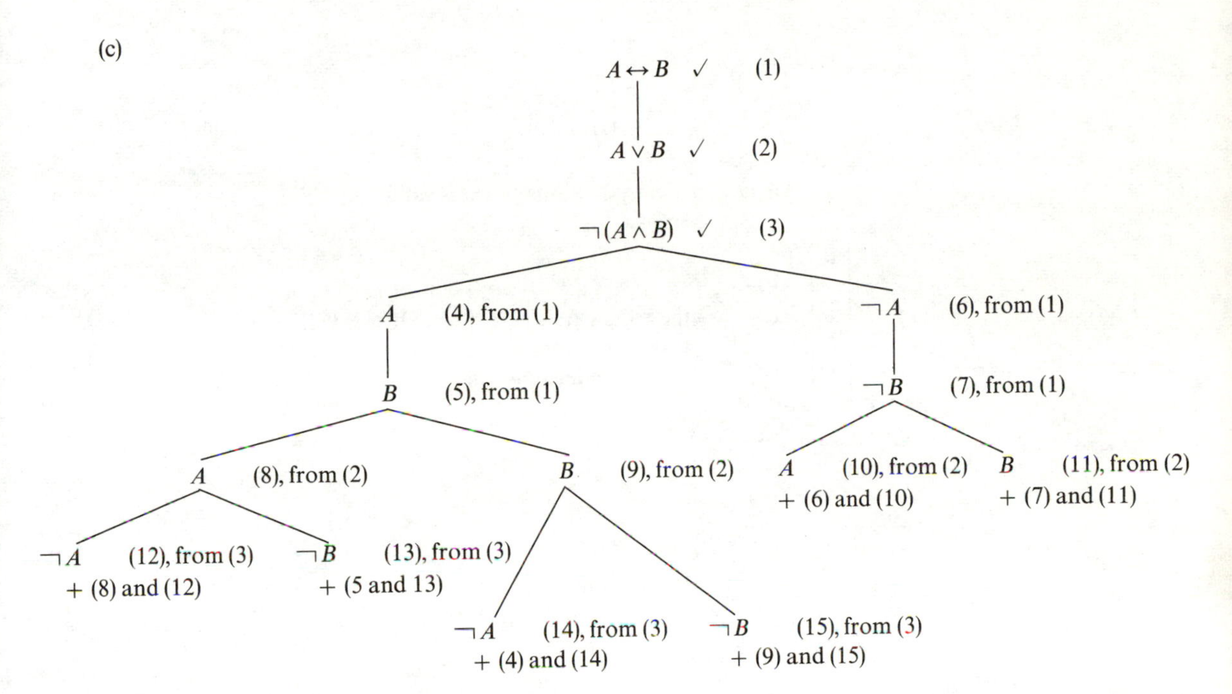

All paths are closed, so **entailment is valid.**

A	B	$A \leftrightarrow B$	$A \vee B$	$A \wedge B$	
t	t	t	t	t	← this is the only line
t	f	f	t	f	where both the
f	t	f	t	f	premises are true,
f	f	t	f	f	so as the conclusion
					is true here too the
					entailment is valid

(d)

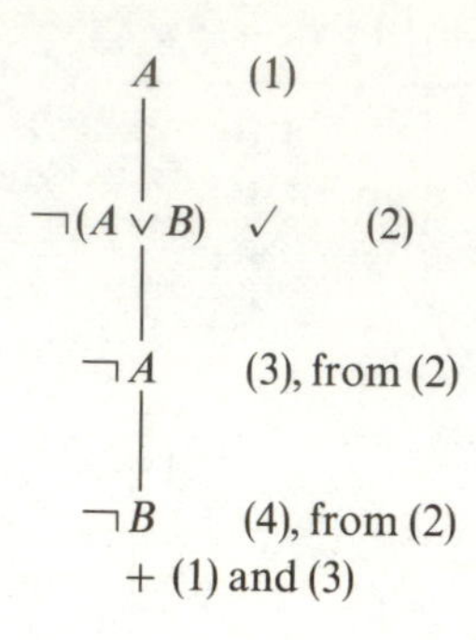

All paths are closed, so **entailment is valid**.

A	B	$A \vee B$	
t	t	t	← these lines are where the premise is true
t	f	t	← and the conclusion is true here too so
f	t	t	the **entailment is valid**
f	f	f	

(e)

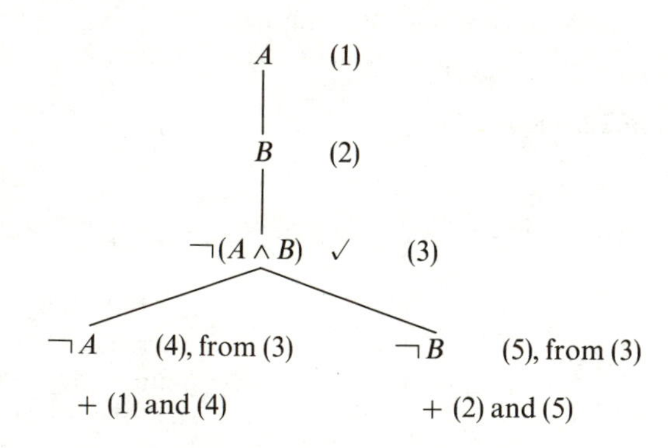

All paths are closed, so **entailment is valid**.

A	B	$A \wedge B$	
t	t	t	← this line is where the premises are both
t	f	f	true and the conclusion is true here
f	t	f	too, so the **entailment is valid**
f	f	f	

(f)

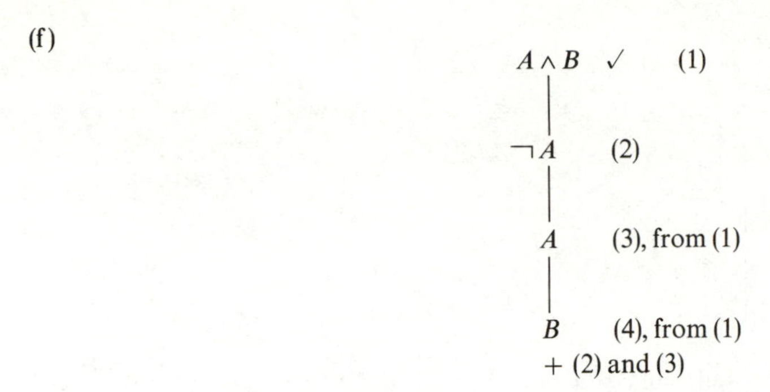

All paths are closed, so **entailment is valid.**

A	B	$A \wedge B$
t	t	t
t	f	f
f	t	f
f	f	f

← this line is where the premise is true and as the conclusion is true too the **entailment is valid**

(g)

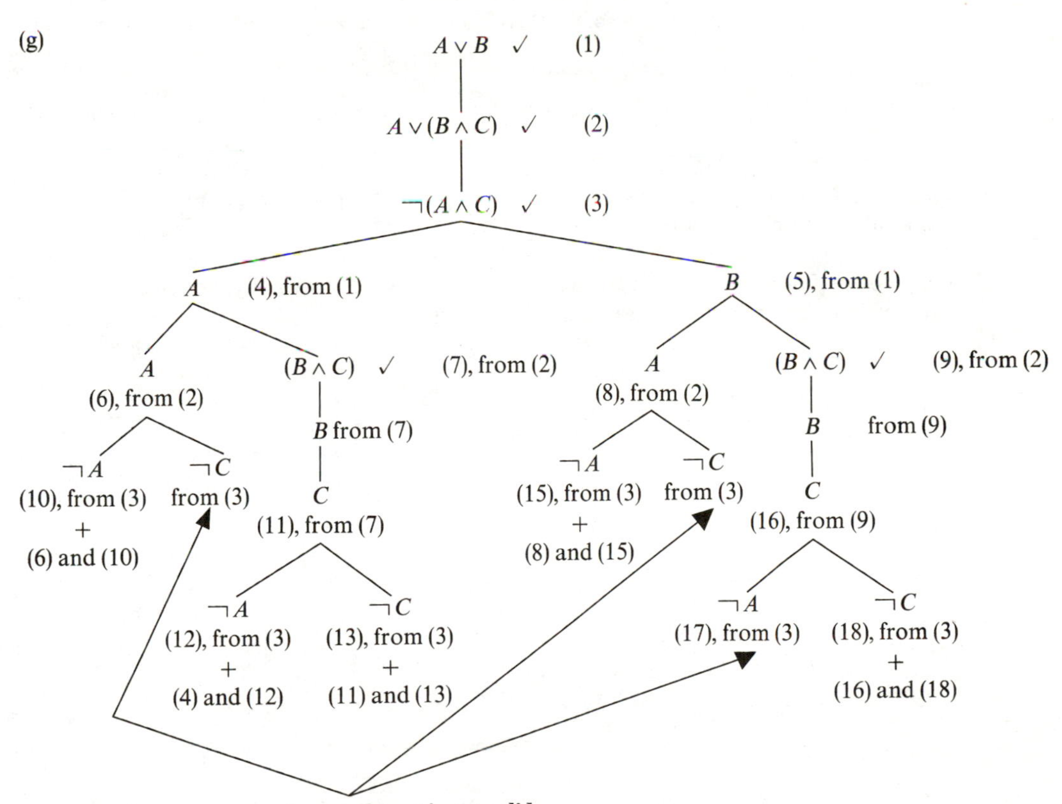

A	B	C	$A \vee B$	$A \vee (B \wedge C)$	$A \wedge C$
t	t	t	t	t	t
t	t	f	t	t	f ←
t	f	t	t	t	t
t	f	f	t	t	f ←
f	t	t	t	t	f ←
f	t	f	t	f	f
f	f	t	f	f	f
f	f	f	f	f	f

The marked lines show that the entailment is not valid.

8.1

(a) There is a possible world in which this can be true.

(b) There is no possible world in which this is true. If it were true in some world then, since $2 + 2 \neq 4$ then there is some number, say n, for which $2 + 2 = n$. Then, if $2 + 2 > n$, by repeatedly subtracting 1 from either side we will eventually get to a stage where we have $m = 0$ where m is not 0. Then, by subtracting $m - 1$ 1s from m we get $1 = 0$ and the argument given in the text just before this exercise leads to a contradiction. A similar argument in the case where $2 + 2 < n$ leads to a contradiction also. So, there is no possible world in which the sentence can be true.

(c) There is no possible world in which this can be true. In any one where it is we get a yard is shorter than a yard, which is contradictory.

(d) There is a possible world where this is true. For instance, one where I am ignorant of, or mistaken about, results on natural numbers.

(e) There is no possible world where this can be true since knowledge is simply the belief of facts, and it is not a fact that there is a greatest prime number.

References

Bibel W. (1987). *Automated Theorem-Proving* 2nd revised edn. Braunschweig: Vieweg

Boolos G. and Jeffrey R. (1980). *Computability and Logic* 2nd edn. Cambridge: Cambridge University Press

Bornat R. (1987). *Programming from First Principles*. Hemel Hempstead: Prentice-Hall International

Bratko I. (1990). *Prolog Programming for Artificial Intelligence*. 2nd edn. Wokingham: Addison-Wesley

Church, A. (1936). A note on the Entscheidungsproblem. *J. Symbolic Logic*, **1**(1), 40–1, 101–2

Clocksin W.F. and Mellish C.S. (1981). *Programming in Prolog*. Berlin: Springer

Fitting M. (1983). *Proof Methods in Modal and Intuitionistic Logics*. Dordrecht: Reidel

Gabbay D.M. (1984). Theoretical foundations for non-monotonic reasoning in expert systems. In *Logics and Models of Concurrent Systems* (Apt K., ed.), pp. 439–59, Berlin: Springer Verlag

Gentzen G. (1934). Investigations into logical deduction. In (1969) *The Collected Papers of Gerhard Gentzen* (Szabo M., ed.). Amsterdam: North-Holland

Gordon M.J., Milner R. and Wadsworth C.P. (1979). *Edinburgh LCF*, Lecture Notes in Computer Science 78. Berlin: Springer

Hamilton A. (1978). *Logic for Mathematicians*. Cambridge: Cambridge University Press

Henson M. (1987). *Elements of Functional Languages*. Oxford: Blackwell

Heyting A. (1956). *Intuitionism: an Introduction*. Amsterdam: North-Holland

Hilbert D. and Ackermann W. (1950). *Principles of Mathematical Logic*. New York: Chelsea Publishing Company

Hodges W. (1977). *Logic*. Harmondsworth: Pelican

Jeffreys R. (1967). *Formal Logic: its Scope and Limits*. New York: McGraw-Hill

Kripke S. (1963). Semantical analysis of modal logic I. Normal modal propositional calculi. *Z. Mathematische Logik und Grundlagen der Mathematik*, **9**, 67–96

Landin P.J. (1964). The mechanical evaluation of expressions. *Computer J.*, **6**(4), 308–20

Landin P.J. (1965). A correspondence between ALGOL 60 and Church's lambda-notation. *Comm. ACM*, **8**, 89–101, 158–65

Landin P.J. (1966). The next 700 programming languages. *Comm. ACM*, **9**(3), 157–66

Lewis C.I. and Langford C.H. (1932). *Symbolic Logic*. New York: Dover

Lloyd J.W. (1984). *Foundations of Logic Programming*. Berlin: Springer

Manna Z. and Pnueli A. (1981). Verification of concurrent programs: the temporal framework. In *Proc. International Summer School on Theoretic Foundations of Programming Methodology*, Marktoberdorf, FRG, June 1981

Manna Z. and Waldinger R. (1985). *The Logical Basis for Computer Programming*. Reading, MA: Addison-Wesley

Martin-Löf P. (1985). Constructive mathematics and computer programming. In *Mathematical Logic and Computer Programming* (Hoare C.A.R. and Shepherdson J.C., eds.). Englewood Cliffs NJ: Prentice-Hall

McCarthy J. (1960). Recursive functions of symbolic expressions and their computation by machine, part 1. *Comm. ACM*, **3**(4), 184–95

Mendelson E. (1987). *Introduction to Mathematical Logic* 3rd edn. California: Wadsworth and Brooks

Pratt V.R. (1976). Semantical considerations on Floyd–Hoare logic. In *Proc. 17th IEEE Symposium on the Foundations of Computer Science*, pp. 109–21

Reeves S.V. (1989). Programming as Constructive Mathematics. In *Proc. of IMA Conference on Mathematical Structures for Software Engineering*, Manchester, July 1988

Reid C. (1970). *Hilbert*. Heidelberg: Springer

Rescher N. and Urquhart A. (1971). *Temporal Logic*. Vienna: Springer

Robinson J.A. (1979). *Logic: Form and Function.* Edinburgh: University Press

Scott D., Bostock D., Forbes G., Issacson D. and Sundholm G. (1981). *Notes on the Formalization of Logic*. Faculty of Philosophy, University of Oxford

Smullyan R. (1984). *What is the name of this book*. Pelican

Smullyan R. (1968). *First-Order Logic*. Berlin: Springer

Sterling L. and Shapiro E. (1986). *The Art of Prolog*. Cambridge MA: MIT Press

Stoy J. (1977). *Denotational Semantics: the Scott–Strachey Approach to Programming Languages*. Cambridge MA: MIT Press

Turing A.M. (1936). On computable numbers, with an application to the Entscheidungsproblem. *Proc. London Mathematical Society, Ser.2*, **42**, 230–65

Wikström Å. (1987). *Functional Programming using Standard ML*. Hemel Hempstead: Prentice-Hall International

Wallen L. (1987). *Automated Proof Search in Non-Classical Logics: Efficient Matrix Proof Methods for Modal and Intuitionistic Logics.* PhD Thesis, University of Edinburgh

Index